BREAKING THE CHAINS
OF GRAVITY

Also available in the Bloomsbury Sigma series:

Sex on Earth by Jules Howard
p53: The Gene that Cracked the Cancer Code by Sue Armstrong
Atoms Under the Floorboards by Chris Woodford
Spirals in Time by Helen Scales
Chilled by Tom Jackson
A is for Arsenic by Kathryn Harkup

BREAKING THE CHAINS OF GRAVITY

THE STORY OF
SPACEFLIGHT BEFORE NASA

Amy Shira Teitel

Bloomsbury Sigma
An imprint of Bloomsbury Publishing Plc

1385 Broadway   50 Bedford Square
New York   London
NY 10018   WC1B 3DP
USA   UK

www.bloomsbury.com

For spaceflight's pioneers who continue to inspire,
and for Mark who believed in me at the start

Contents

Preface   8

Chapter 1: Hobby Rocketeers   11

Chapter 2: The Rocket Loophole   27

Chapter 3: The Turning Tide of War   53

Chapter 4: Escape and Surrender   71

Chapter 5: Nazi Rockets in New Mexico   89

Chapter 6: Rockets Meet Airplanes   107

Chapter 7: A New War, a New Missile, and a New Leader   123

Chapter 8: Higher and Faster   133

Chapter 9: Edging into Hypersonics   155

Chapter 10: The Floating Astronaut   171

Chapter 11: Space Becomes an Option   191

Chapter 12: The First Satellite Race   211

Chapter 13: One Little Ball's Big Impact   225

Chapter 14: The Fight to Control Space   251

Epilogue: America Finds Its Footing in Space   267

Glossary of People   271

Glossary of Places and Organizations   273

Glossary of Rockets   275

Selected Notes   277

Bibliography   290

Acknowledgments   297

Index   299

Preface

Popular retellings of the National Aeronautics and Space Administration's history typically follow the same narrative: In 1961, President John F. Kennedy pledged the nation would land a man on the Moon by the end of the decade and return him safely to the Earth. In July of 1969, Neil Armstrong took one small step on the Sea of Tranquillity, fulfilling the fallen president's dream and completing a technologically daunting task in the name of restoring America's national prestige. It was, by all accounts, a remarkable achievement given spaceflight's embryonic state in the 1960s and the short time frame. The story becomes even more incredible in light of the fact that NASA was just three years old and hadn't yet put a man in orbit when Kennedy promised America the Moon. Though common, this version of the story creates the illusion that NASA invented a lunar landing program in response to a presidential decree.

NASA wasn't created in a vacuum and suddenly tasked with the Moon landing. The agency might have been incepted in 1958, but it was assembled from preexisting parts, drawing off decades of research in rocketry, human tolerances, hypersonic flight, and the bureaucracy needed to oversee a major undertaking like a lunar landing program. NASA has technological and bureaucratic roots stretching back decades before it formally opened for business that made the Apollo program possible, and these roots are what this book is about.

The incredible rockets that launched America's first astronauts reached the nation by way of German engineers imported after the Second World War and employed by the U.S. military. The knowledge of human survival in space came from early air force programs, some done with primates and some with humans. Knowledge of how a vehicle could return safely from orbit was largely the product of the National Advisory Committee for Aeronautics, the agency that was also versed in bridging the gap between military and civilian partners in cutting-edge aeronautics programs.

But it must be said that this book only tells part of the story. Almost every rocket, aircraft, person, organization, and research laboratory in this book merits a work dedicated to its history. In fact, most of them do have dedicated volumes. In simplifying the story to bring it to a broader audience, I decided to focus on certain people and follow certain narratives to the exclusion of some notable figures like American engineer Robert Goddard and Russian scientist Konstantin Tsiolkovsky, two fathers of rocketry whose contributions to the field were invaluable. The complete, unedited story would be a tome that only die-hard space fans would have the patience to sift through. Spaceflight is part of our shared human history and shouldn't be an opus accessible only to initiates. It should be available to everyone interested in exploring this rich history.

Among other decisions I made in writing this book was the decision to use male pronouns. With a handful of exceptions, everyone working in aviation and aeronautics between 1930 and 1958 was male, and most forward thinkers assumed that the first person in space would be a man. It is a thought process indicative of the era. I also chose to keep center names and dollar amounts consistent with the time frame of the book. Names are not reflective of current monikers, and values have not been adjusted for inflation.

My hope is that this book opens up NASA's prehistory to those who might not realize that America's national space agency has such a fascinating backstory, and that it inspires a few to dig into this history a little further. Humanity's exploration of space is wonderful. Having a deeper understanding of how it all started is not only interesting, having a sense of the context makes everything we have achieved in the last half century of space exploration that much more incredible.

CHAPTER ONE

Hobby Rocketeers

On May 17, 1930, dusk fell just before nine o'clock at the end of a warm, clear Saturday in Britz, Berlin, but Max Valier showed no signs of leaving his workbench for an evening of leisure. He remained in his seat, focused on a simple combustion chamber bolted to the table in front of him. It was a simple setup. At the center was a combustion chamber, a simple steel tube with an upward-facing exhaust nozzle. On the other end were a series of small bore holes through which the fuel and oxidizer were introduced. The whole apparatus was set up on a grocery scale. His assistants Arthur Rudolph and Walter Riedel sat some distance away at two tanks, one of kerosene mixed with water and the other of liquid oxygen. The two men manually opened the valves as Valier dictated, sending the fuel and oxidizer into the combustion chamber where they mixed. Once the combustion chamber was adequately pressurized, Valier lit the mixture with a blowtorch. As the jet of flaming gases roared upward from the combustion chamber, directed by the nozzle, the resulting reaction was a downward force onto the scale. While the engine burned, Valier added weights to the other side of the scale until it was properly balanced, giving him a crude measure of the engine's efficiency.

That day, Valier had made two successful tests with the same setup. Two good burns in the combustion chamber had yielded good data. A third test had failed, the accompanying jolting motions deforming the test hardware at the same time. At that point Riedel had pushed for the skeleton crew to end their day and start fresh the next day, but Valier's enthusiasm had been indomitable. He was so encouraged by the afternoon's successes that he pushed for one final test to end the day on a high note. The combustion chamber was reassembled, the fuel and liquid oxygen tanks were hooked up.

As he had done dozens of times before, Valier moved the flame toward the pressurized combustion chamber, but instead of the slow, steady burn he was expecting, the space was rocked by an earsplitting explosion. Unbeknownst to Valier, part of the emulsified mix of kerosene and water had combined with liquid oxygen to form a jellylike substance that stuck to the sides of the chamber where it burned with explosive, deadly force. Riedel instinctively closed the valves on the tanks and rushed to Valier's side, barely catching the man before he collapsed to the floor. Passing a stricken Valier off to Rudolph and their machinist, Riedel ran outside in search of a passing car to send for help, but it was too late. There had been no protective shield between Valier and the combustion chamber. When it exploded, a piece of shrapnel had pierced Valier's pulmonary artery. That day, one of Germany's most notable popularizers of rocketry and space travel bled to death on the floor of a Spartan laboratory.

Valier had made a name for himself and for rocketry by undertaking similarly dangerous and almost foolhardy public experiments extolling the virtues of rocket propulsion. It was a boyhood fascination come to life, full-scale experiments by the man who, as a child, had attached firecrackers to model airplanes and sent them hurtling through the skies of Innsbruck in Austria during school holidays. As an adult, Valier found a kindred spirit in Romanian-born physicist Hermann Oberth.

Oberth found rockets through French novelist Jules Verne. While recovering from a bout of scarlet fever in Italy when he was fourteen, Oberth read Verne's 1865 novel *De la Terre à la Lune (From the Earth to the Moon)*, which tells the story of a group of Americans from the Baltimore Gun Club who build a massive cannon and shoot themselves in a train-like vehicle to the Moon. More than the fantastic story, it was the realistic potential of rocket propulsion for spaceflight that had captivated Oberth, though he knew black powder like the characters in the story used couldn't get a spacecraft to the Moon. Black-powder rockets simply didn't have enough power. But Oberth suspected liquid-propelled rockets would.

And so he set out designing a simple, proof-of-concept rocket called a recoil rocket that would propel itself through space by expelling exhaust gases from its rear end. It was a basic application of Newton's third law that states that every action has an equal and opposite reaction. The expulsion of gas behind the rocket would propel it forward. But a paternal tradition intervened to derail Oberth's pursuit of rocketry. In 1912, he moved to Germany to study medicine at the University of Munich. It was a short-lived career. After serving in the First World War with a medical unit on the battlefield, Oberth determined he was not destined for life as a physician.

After the war, Oberth returned to the university, resolved to change his path. He switched his field of study from medicine to mathematics and physics, and self-specialized in rocket propulsion. The work culminated in a doctoral dissertation on liquid rockets and their application for spaceflight, but the work was rejected by his advisers in 1922. Though astonishing, his committee said, the paper failed to meet the requirements for a degree in classical physics. Rocketry and spaceflight were fodder for science fiction, they believed. It was a blow strong enough to turn the young physicist away from academia but not from the pursuits of rocketry. Oberth circulated his rejected thesis among publishers and eventually found a small press willing to print the volume.

Die Rakete zu den Planetenräumen (The Rocket into Planetary Space) reached bookshelves in 1923, but it wasn't widely well received. Though less than a hundred pages long, it was sufficiently dense and loaded with complex diagrams and calculations to alienate the casual reader. But the book did strike a chord with amateur rocket enthusiasts who were similarly taken by the prospect of rocket propulsion as a means to space travel. Among *Die Rakete*'s avid readers was Max Valier, who was so taken with the work that he wrote to Oberth in 1924. The initial letter sparked a fruitful correspondence. Oberth acted as the teacher to the enthusiastic student Valier, the pair discussing the fundamentals of rocketry, best practices of testing, and even a plan to publish a book together. The joint work would play to both men's strengths. Oberth would supply the technical details while Valier would finesse the writing such that it would be accessible to the layman. But it wasn't long before the duo reached an impasse regarding methods. In discussing plans for future rocket research, Oberth wanted to undertake a step-by-step test program to gradually explore and understand the power of liquid-propelled rockets, while Valier wanted to use available powder rockets to gather basic data points. Valier also wanted to carry out these experiments in public, something Oberth found flashy and unscientific. But Valier knew this was the best way to secure patrons for an undertaking as lofty as rocket research.

Valier had subsidized a career of research and public talks by strapping rockets to anything that moved, beginning with cars. Valier found a willing patron in German automotive industrialist Fritz von Opel, who was willing to strap rockets to one of his vehicles as a way of demonstrating their power. The first collaboration between the automobile manufacturer and the rocket enthusiast was the Opel-Sander Rak.1, a standard Opel race car whose engine had been removed. Its new power source was a cluster of solid rockets manufactured by Friedrich Wilhelm Sander and strapped to the rear. But it hardly looked like a rocket car. Boxy in the front with wheels sitting on either side of its main body, the Opel-Sander Rak.1 was not an aerodynamic streamlined design. It was, however, ready for a test drive.

On March 12, 1928, the Opel-Sander Rak.1 was parked on the circular racetrack at the Opel factory. Behind the wheel sat race car driver Kurt C. Volkhart. No one knew exactly what to expect, but with his life on the line even a seasoned driver like Volkhart sat with his hands stretched out before him, bracing himself like he was about to be shot from a cannon. The rockets were lit before a small crowd of onlookers waiting anxiously to either see a rocket car race down a track or erupt in a fiery explosion. Two bright jets appeared amid a cloud of smoke, signaling that the fuses were burning. When the fire reached the rockets, there was a sudden loud hissing noise. The Rak.1 started moving, but almost before it could gather momentum the powder was exhausted and the rockets lost thrust. The world's first rocket car had covered five hundred feet in thirty-five seconds with a top speed of just five miles per hour. On the sidelines, von Opel failed in his attempts not to laugh. Unwilling to stand by and be ridiculed, Valier decided to sacrifice one of his small, two-inch bore rockets. Without an aerodynamic cap or the right length guiding pole, he launched the rocket. The gathered spectators, including von Opel, watched as it shot to a height of more than thirteen hundred feet in about two seconds. This simple demonstration silenced von Opel's laughter and brought the media around to the exciting potential of rockets applied to travel.

Valier knew the problem with the Rak.1 came down to friction; the car's wheels moving against the asphalt was too much for the rockets to overcome. And so he tried taking advantage of momentum, lighting the rockets when the car was already moving. The results were better, though still not the explosive run he wanted. With the car traveling at eighteen miles an hour, the lit rockets increased its speed threefold. Building off this first successful run, Valier and von Opel developed a second version of the rocket car, the Opel Rak.2. It was a more streamlined design: Its forward end was tapered like a bullet, and wheels were housed in wells within the body. Valier even added small inverse wings on either side of the car to keep the wheels pressed to the ground, just in case the rockets accelerated the car enough for it to become airborne. The Rak.2 was also designed to allow the driver to ignite the rear-mounted rockets by a pedal in his footwell. With twenty-four rockets in the back and von Opel himself at the wheel, this second rocket car reached a top speed of about 145 miles per hour on the Avus Speedway in Berlin two months after the first Rak.1 failed to impress. It was a vast improvement over previous rocket runs, but it didn't surpass the power of traditional combustion engines. Weeks before the Rak.2's run, American driver Ray Keech set a landspeed record of nearly 208 miles per hour at Florida's Daytona

Beach Road Course in the triple-engined internal combustion White Triplex Spirit of Elkdom.

Asphalt, Valier realized, was still a problem, creating too much friction with the tires for the rocket cars to reach their top potential speed. To solve the problem, Valier developed a vehicle designed to run on rails. The Eisfeld-Valier Rak 1, so named to reflect a new partnership with the J. F. Eisfeld Powder and Pyrotechnical Works firm from Silberhütte-Anhalt in central Germany, made its first test runs in the Harz Mountains at the end of the summer and early autumn. Valier's instinct might have been right, but the execution revealed a host of problems with this design. The wheeled sled gained too much speed, flew off the track, and crashed. The wheel struts mounted below the vehicle broke and destroyed tracks. One test in early October saw the wheels completely separate from the vehicle. It was an embarrassing failure for Valier, who had invited guests from the national railways to the October test in an attempt to secure a new investor for his idea. Instead, the incident prompted local authorities to step in. His rocket-powered vehicles were banned as safety hazards, though the ban was eventually lifted.

The failed rail runs inspired Valier to try his hand at developing a rocket vehicle with no moving parts. What emerged from this goal was a long, slender sled sitting on skids with a rear seat for a pilot. Behind the seat was a bank of small rockets. The Rak Bob, as it was called, made a series of demonstration runs at a winter sports festival in early 1929 on Bavaria's Eibsee Lake. It was one of Valier's more promising designs; so great was his faith in the vehicle that he felt safe enough to let his wife, Hedwig, behind the wheel. The sled tore across the wintry landscape at impressive speeds, traveling so fast that when Valier studied the tracks of some test runs, he found they disappeared in places. The Rak Bob actually went fast enough to lift off the ground. But fast land vehicles had never been Valier's goal. What he had always wanted to develop was a means of achieving rocket-powered flight, and for that his only means for testing was to attach rockets to lightweight airplanes.

The Wasserkuppe is a high plateau in Germany's Rhön Mountains. In the 1920s, it was a popular spot for sailplane pilots, who rode the strong updrafts high above and across the valleys stretching out below. In March 1928, Valier and Sander took a trip to the Wasserkuppe to meet with sailplane designer Alexander Lippisch. Valier didn't tell Lippisch his name or his intentions in the meeting, maintaining a mysteriously low profile while peppering Lippisch with questions about a custom sailplane design. He wanted something lightweight and tailless for a very special purpose, he told the sailplane designer, divulging only that it would involve a high thrust rear-mounted engine. Lippisch and his designers were skeptical of the stranger's odd request but consented to build the plane. Days later, Lippisch recognized Valier in a picture accompanying an article in a local newspaper about rocket cars. Realizing who the stranger was, Lippisch became more interested to see how the rocket popularizer's mysterious sailplane test would turn out.

The sailplane Valier eventually procured was a specially modified Lippisch Ente (which means "duck" in German), a moniker derived from the plane's long, beak-like structure jutting out front. In place of the absent tail were two cylindrical rockets packed with nearly nine pounds of compressed black powder that were wired to be electrically ignited from the pilot's seat. In the cockpit on the morning of June 11, 1928, was one of Lippisch's company test pilots named Friedrich Stamer. As Valier had done with the rocket cars, Stamer got the Ente flying before lighting the rockets one at a time. A quiet hissing told him the first rocket was burning; for the moment there was no significant speed increase to disrupt the easy flight. Seconds later, the hiss was replaced by a booming explosion. In an instant the Ente was on fire. Without succumbing to panic, Stamer managed one of the finest landings of his career to that point, getting the flaming plane to the ground before the second rocket had a chance to ignite. Only once he was on the ground did the pilot react with panic. He wriggled out of the plane and rolled in the wet grass to extinguish and soothe his own rear end while the Ente's burned. Stamer walked away from the flight unharmed. Unfortunately, the same could not be said for the plane.

Years of mixed results in testing rocket-powered automobiles, sleds, and sailplanes all fed into Valier's ultimate dream of using rocketry to shrink the world. He imagined a future where people would cross the Atlantic Ocean from Europe to America in under an hour thanks to rocket propulsion. And from there, he pictured rockets propelling planes higher and faster to the point where they could escape the Earth's atmosphere to fly in space. But he knew this dream could never be a reality if he didn't move away from the black powder rockets he had been using in his public demonstration tests. Powder simply couldn't deliver the power needed to send an airplane rocketing across an ocean or into space. For that, he would need liquid propulsion, and he sought to bring more minds to bear on the challenge of developing this science.

In the back parlor of an alehouse in Breslau on June 5, 1927, Valier met with a small cohort of rocket enthusiasts, scientists who also wanted to bring this fantastical idea firmly into the realm of reality. The group, all drawn to the rockets described in Oberth's *Die Rakete*, founded the Verein für Raumschiffahrt (Society for Space Travel) that afternoon with the goal of actually building the rockets described by their icon. Their ultimate goal was summed up in their simple motto: "Help to create the spaceship!"

Oberth joined the VfR in 1929, following an invitation from Valier, one of the last letters between the two men before disagreements over how to approach rocket testing led to their falling out. By then, the society's roster had swelled to more than eight hundred members, from writers to engineers to scientists. And many, like Oberth, found Valier's methods more frustrating than fruitful. Though the popularity of his experiments was certainly good for exposure and patronage, most felt Valier's showmanship denigrated what they were trying to do, turning rocketry into a sideshow rather than a serious field of science. Though he was a founding member, a rift soon developed between Valier and the VfR. It widened until rocketry's greatest cheerleader left the society he'd founded.

The same year Oberth joined the VfR he released an expanded version of his published thesis. The new book, retitled *Ways to Spaceflight*, was more than four hundred pages long and secured Oberth's place as a legitimate rocket scientist. This edition was far more readable, using stories of flights to the Moon to illustrate difficult concepts, and expanded discussions that answered questions critics had raised after reading the earlier edition. The new book also brought six years of additional research to the discussion, covering topics such as optimal flight trajectories for rockets carrying payloads off the planet. Launching a rocket straight up was instinctual but inefficient, Oberth wrote. It was similar to what Valier found with his first rocket car experiments, only this time it was friction with the air and not asphalt that would hinder the rocket. When traveling directly through the atmosphere, the air would drag on the rocket, costing precious energy. It would be better, Oberth suggested, to launch a rocket on an easterly path, allowing it to use the centrifugal force of the Earth's rotation for a slight increase in speed that would make up for the speed lost through atmospheric friction.

Oberth's expanded work reached well beyond his scientific peers and into the realm of artists, catching the attention of Fritz Lang. The acclaimed silent film director had debuted his eleventh directorial effort, *Metropolis*, four years earlier to great fanfare. The film, which tells the grisly story of a futuristic world where subterranean slave labor powers the lives of the rich who live aboveground, entranced audiences. Building off that success, Lang sought to capitalize off the national interest in rocketry. Lang's wife, Thea von Harbou, had written a script about an intrepid crew who fly to the Moon to confirm an aging doctor's theory that its mountains hold more gold than any reserves on Earth. In bringing von Harbou's story to life, Lang wanted the film, *Frau im Mond (Woman in the Moon)*, to balance the fantastic plot with technical accuracy. And so he hired Oberth as the film's technical adviser.

The prospect delighted Oberth; he knew a realistic movie about rockets would not only popularize the field, it might bring in money for research as well. The physicist threw himself into the project, advising Lang on practical aspects like the size and architecture of the rocket as well as a realistic depiction of the best trajectory for a flight to the Moon. Though Lang exerted his artistic license over Oberth's technical direction, the film was nevertheless the first to depict a realistic rocket leaving the Earth. In the launch scene, Lang shows the massive rocket crawling along rails from its assembly building to the launch site while searchlights sweep over its impressive structure. Once at its destination, the rocket is lowered into a tank of water, covering it almost to the nose—a means to absorb the vibrations and acoustics of the fiery launch. The rocket's engines ignite, bubbling the water away as it shoots upward toward the Moon to the delight of the gathered crowd watching from nearby bleachers.

Though Oberth was certainly pleased to have a hand in advising the technical aspects of *Frau im Mond*, it was a spin-off from this project that really excited the scientist. Willy Ley, a fellow member of the VfR and avid rocket popularizer, suggested that Oberth build and launch a real liquid-fueled rocket to coincide with the film's release. It could not be a rocket large enough to go to the Moon as in the film, but it would be an incredible publicity stunt. The film company, Universum Film-Aktien Gesellschaft, and its marketing department backed the idea and granted Oberth a small amount of funding. There was, however, one condition. UFA stipulated that Oberth had to pay the company 50 percent of any future proceeds he might make from the technology he developed in building this rocket.

Potential future profits weren't on Oberth's mind when he brought fellow VfR member Rudolf Nebel on board and began designing his rocket. Its power source was a liquid-fueled engine of his own design that burned a mixture of liquid oxygen and gasoline. But turning his theory, calculations, and thought experiments into a viable flying rocket turned out to be far easier said than done. His choice of fuel turned out to be a tricky mix. One propulsion experiment had him observe the behavior of a fine stream of gasoline as it was introduced into a container filled with liquid oxygen. The mixture was lit as it would be inside the rocket's combustion chamber, but the reaction became explosive. The blast burst Oberth's eardrums and burned the skin around his left eye, nearly costing him the organ. Gasoline and liquid oxygen, the physicist learned that day, combust considerably faster in a limited, narrow space.

Unfortunately, Oberth's progress was slower than he had anticipated, and he couldn't meet UFA's planned launch day of October 29, 1929. The company issued a simple press release in advance of *Frau im Mond*'s premiere explaining that the launch had been canceled due to unfavorable weather. It was an entirely fabricated excuse, but the public wouldn't know the difference, and it saved Oberth a considerable amount of embarrassment, though he was not immune to the sting of failure. The film was incredibly well received at its October 15 premiere, but while the social, intellectual, and political elite praised Lang for creating another masterpiece, Oberth was all but ignored. Disillusioned and broke, he left Berlin for Romania to rejoin his family.

A few months later, a new opportunity arose to lure Oberth back to Germany. Nebel had secured funding from the government-sponsored Reich Institute for Chemistry and Technology, and with this influx of money the VfR hoped to bring Oberth's rocket to life. Oberth consented to return after Nebel's persistent imploring invitations, and this time he set to work with a larger team plucked from the VfR. He recruited Klaus Riedel, an engineer from the Siemens Company, which manufactured electronics for radios and electron microscopes, along with Wernher von Braun, a teenage engineering student at the Berlin Institute of Technology. An avid amateur astronomer fascinated with the stars and the idea of visiting other planets, von Braun had also become a fan of Oberth's after reading the original *Die Rakete* in 1923. The young student saw in Oberth's work a means to leave the planet, and was in the rare position of having the means to engage in this costly pastime, afforded him by his wealthy father, Magnus Freiherr von Braun.

Valier had given his life to the pursuit of liquid propulsion when Oberth's new team began experiments that were strikingly similar to the one that claimed their former colleague. At the core of the rocket they were trying to build was a basic engine Oberth called the Kegeldüse. Its combustion chamber was a hollow steel cone with two inlet ports, one for the fuel and the other for the oxidizer. To test the engine, the team secured the chamber inside a metal bucket filled with water as a coolant, oriented so its exhaust end was facing up and any thrust produced pushed downward. The bucket was then placed on a grocer's scale, allowing Oberth, like Valier before him, to measure the thrust of the engine.

Oberth's first test came on July 23, 1930. Rain cast a gloomy mood over the group, exacerbated by the presence of an outsider. The director of the sponsoring Reich Institute for Chemistry and Technology was on hand to verify the results and confirm that the money invested hadn't been wasted. Oberth anxiously watched as his assistants set up the test, then pressurized the combustion chamber. It fell to Riedel, as one of the junior members on the small team, to throw the burning, gasoline-soaked rag into the upward-facing exhaust nozzle. He did, and there was no explosion. Instead, a roaring three-foot-tall spear of flame shot upward from the chamber. The Kegeldüse burned for ninety seconds and delivered a constant thrust of 15.4 pounds. The success of the test vindicated Oberth after the *Frau im Mond* failure, but without additional research funding he was again forced to return to Romania where a job teaching math to high school students would allow him to support his family. The rocket research stayed in Germany in the hands of Nebel, von Braun, and the rest of the VfR.

The VfR, however, did not have the means to continue where Oberth's testing left off. They were without university, military, or industry sponsorship and also without a dedicated laboratory space. In addition, they faced legal restrictions over rocket testing, a frustrating if understandable result of Valier's death. Nebel and Riedel managed some small experiments at a family farm in the latter's hometown, but it was clear they needed a better space to work, some permanent facility to serve as their home base. If the VfR was going to crack the secrets of liquid-fueled rocket propulsion, they would need a better test site to work the kinks out of their designs than a field in Riedel's native Bernstadt.

It was Nebel who found the group a home in the fall of 1930. Down a bad road in Berlin's northern suburb of Reinickendorf, he found a vacant property surrounded by a wire fence. Covering roughly two square miles, the site consisted of a half dozen munitions storage buildings, each surrounded by earthen walls forty feet high and sixty feet thick with narrow passages giving access to the buildings. It was a former ammunitions dump disused since the First World War and a perfect site for rocket testing. Nebel negotiated a lease with the city, which granted the VfR access to the munitions storage sites as well as a small administrative building on the condition that no facility be permanently altered and no equipment installed that could not be removed within forty-eight hours. The cost of the lease more than made up for the restrictive terms. The VfR leased the site for one year for just ten German marks ($42 or £204). On September 27, 1930, the day the group took possession, Nebel mounted a sign on the front that read RAKETENFLUGPLATZ BERLIN. It was Berlin's first rocket airdrome.

The Raketenflugplatz slowly took shape. The first members of the VfR to arrive set up Spartan living quarters and laboratories before turning their attention to the task of developing a viable liquid-fueled rocket. Without a sponsor, the group relied on members with financial means donating what they could toward building test stands and parts for rocket engines, but more often the group turned to bartering. They would trade unneeded materials for things they did need and offer their skills as mechanics for parts. At times they even bartered for food to keep the team going.

Taking a leaf out of Valier's book, the VfR also sought to engage the locals and secure donations through public demonstrations. Nebel took the lead in publicizing the VfR's activities, inviting visitors to the Raketenflugplatz for engine tests. Unfortunately, funding and donations typically hinged on a successful demonstration, and with success fairly rare, the VfR was forced to proceed with minimal resources and simplistic test setups. But when success came it was exciting. The team was testing a Repulsor rocket in early 1931, a design consisting of an engine encased in water for cooling with a long support stick training behind it for guidance. The stick was connected to the liquid oxygen and gasoline tanks that fed the engine. Four fins attached to the base were designed to keep it from flying erratically. A parachute in a compartment at the rocket's rear end would deploy to slow its landing after the powered flight was over. All told, it was a small rocket, about twelve feet long and weighing about forty-five pounds when fueled. On February 21, the team was testing the Repulsor's engine when the whole rocket suddenly shot into the air. It rose to about sixty feet before falling back down, sustaining minor damage in the process. It was an accidental first launch, but one the VfR was very pleased with.

The first year at the Raketenflugplatz was a busy one, with 87 launches and 270 static firings, engine tests with the hardware bolted firmly in place so it wouldn't move. And things seemed poised to improve. The public was still interested in rockets in 1931, so much so that the media was seeking out interesting rocket tests for the sake of a good story. In October, Universum Film-Aktien Gesellschaft, the film company behind Lang's *Frau im Mond*, sent a crew to the VfR's suburban facility to film a weekly newsreel featuring a Repulsor launch. With the cameras rolling, a rocket roared to life and cleared the launchpad, disappearing into the sky before the delighted film crew's eyes. Soon after, the parachute tore off, and the rocket, still sputtering flames from the remnants of the gasoline in its fuel tank, weaved through the sky before landing on an old shack across the road from the Raketenflugplatz.

The last traces of gasoline in the tank set fire to the structure, which turned out to belong to the local police. Uniformed officers stormed the Raketenflugplatz, demanding all rocket testing be stopped immediately. The VfR fought back, arguing that the odd test was bound to go wrong. After much discussion, tempers cooled, and the men settled on a compromise: The VfR could continue firing their rockets provided they place firmer safety restrictions on their activities. It was a happy resolution for the VfR and an entertaining newsreel for UFA. But the media wasn't the only group with money that was starting to take notice of the rockets flying about the Raketenflugplatz.

CHAPTER TWO

The Rocket Loophole

On a spring day in 1932, three men arrived at the Raketenflugplatz. Though dressed in ordinary street clothes, their demeanors hinted that they weren't simply interested passersby. The leader of the small group was Colonel Karl Becker, the German Army's chief of Ballistics and Ammunition. Accompanying him was his ammunitions expert, Major Ritter von Horstig. The third man was Captain Walter Dornberger, a lifelong military man whose sole break from the service had come in 1925 when he left briefly to study engineering at the Berlin-Charlottenburg Technische Universität. He resumed service with the army in 1930 with both a bachelor's and a master's degree in hand, credentials that facilitated his transfer to the ballistics branch, where he began studying the potential of rockets as offensive weapons. On the day he arrived at the Raketenflugplatz, Dornberger was chief of the army's powder rockets development program.

The German Army's interest in rockets as weapons was fairly unique. The offensive use of rockets had fallen out of favor decades earlier with the advent of better artillery and combat aircraft, new technologies that quickly replaced hand-packed cannons that could not be precisely aimed. Aviation also matured rapidly during the First World War. The first airplanes were used as reconnaissance aides, flying in front of cavalry and ground troops for a bird's-eye view of enemy armies lying in wait. The rear-seated navigators in these reconnaissance planes began carrying guns to fire on enemy armies but soon found themselves firing against their cohorts in the air. To simplify the increasingly common aerial gunfights, armies on both sides of the conflict mounted guns to the front of their airplanes, allowing pilots to direct their fire by flying toward their targets. It was the Fokker aircraft company that made the most significant advance to aerial warfare. Fokker perfected an interrupter gear that ensured automatic weapons always fired between the spinning propeller blades, decreasing the chances of a bullet hitting the spinning blades and ricocheting off in an unwanted direction. Combining the interrupter gear with the German innovation of all-metal fuselages made the nation's air force one to be reckoned with and helped usher in the era of aerial dogfighting.

Developments in combat aircraft had not won World War I for Germany. The conflict left the country devastated, and the Treaty of Versailles, which had enforced the war's end, had left the German military in ruins. In an attempt to prevent the vanquished nation from rearming itself and instigating a new conflict, the treaty stipulated that the German Army be limited to seven divisions of infantry and three divisions of cavalry, totaling four thousand men with no more than three hundred in leadership positions. The German air force was restricted even further. The treaty ordered all flying personnel demobilized and limited the future air force to a thousand men. Completed airplanes and seaplanes; dirigibles and other lighter-than-air vehicles; all vehicles under construction; and all the supporting infrastructure, including plants manufacturing the lifting gas hydrogen, airplane parts, and all munitions were forcibly surrendered to the governments of the Principal Allied and Associated Powers. The Treaty of Versailles failed to address rockets simply because they had not been used offensively in the First World War. Though forcing Germany's air force into submission severely weakened the nation's overall military strength, the omission of rockets opened a loophole. The German military could revisit this technology, rearming itself and developing a long-range bombardment capability, without blatantly violating the terms of the cease-fire.

It was with this intention in mind that Becker, von Horstig, and Dornberger arrived at the Raketenflugplatz that spring day, though they had been invited by Rudolf Nebel. In seeking a source of funding for the VfR, Nebel had hand delivered a copy of his technical treatise "Confidential Memo on Long-Range Rocket Artillery" to Becker. It wasn't a scientifically perfect paper, but the concepts were sufficiently well developed and intriguing that Becker was moved to take a trip to the outskirts of Berlin to see what these young men were working on.

There were two simple rockets undergoing testing at the Raketenflugplatz that day, the Mirak 1 and the Mirak 2. The moniker Mirak was an abbreviation of *minimumrakete*, meaning "simple rocket." The Mirak 1 looked like a firecracker. It had a simple copper rocket engine, a smaller version of Oberth's Kegeldüse engine, housed inside a cylindrical fuselage, nestled behind the bullet-shaped cover with a long aluminum tube that trailed behind it as a guiding stick. Though this was the rocket ready for a launch, the army delegates were more interested in the larger and more sophisticated Mirak 2. Without one to launch, the VfR displayed their instrumentation and data from earlier tests of the advanced rocket before demonstrating its engine's power with a static fire test.

The army representatives weren't impressed with the VfR. To men used to military precision, the group's poor record keeping and a lackluster engine test gave the impression that the VfR's greatest accomplishments lay in flashy shows of explosives. But Becker could see there was talent in the group and offered the VfR a chance to build an advanced rocket with the army's sponsorship and launch it from the military site at Kummersdorf West, some seventeen miles from Berlin. A successful launch, hinted Becker, could lead to the army becoming a benefactor for the VfR.

Months passed while the VfR worked on their new rocket, and at the end of July it was ready to launch. Early one morning before sunrise, two cars drove out of the Raketenflugplatz on their way to Kummersdorf West. The first car carried the rocket, a one-stick Repulsor modified from the original design to meet the army's specifications. The second car carried liquid oxygen, gasoline, and all the tools needed to erect and launch the rocket. Also in the cars were the men who would have the honor of launching the

morning's test: von Braun, Nebel, and Riedel. Dornberger
joined the men as a representative of the army.

The sophistication of Kummersdorf West awed the VfR
pioneers. They had never seen so much measuring equipment
in one place, everything from cameras to precision time-
keepers like chronographs and optical tracking instruments
like phototheodolites. These were the tools that could gather
exact data on their rocket, presuming it took flight. The men
set to work preparing their Repulsor for launch, and by two
o'clock that afternoon everything was ready to go. At the
signal, the rocket leaped into the air, rising about one hundred
feet before tipping over horizontally on a flight path that sent
it crashing into the forest. Worse still, they hadn't managed to
capture any data. There would be no money from the army
to fund the VfR's ongoing work. But while the rocket failed
to impress the German Army, Wernher von Braun did leave
a mark on Dornberger. Engaging the young, fair-haired
engineer in conversation about past and future tests and technologies, Dornberger was struck by von Braun's shrewdness
and technical knowledge of rocketry. Von Braun, he could
tell, held great promise for the world of rocketry.

Dornberger might have added determination to the list of
von Braun's appealing qualities. Unwilling to let the
Repulsor's failure at Kummersdorf West be the end of a
potential partnership between the army and the VfR, von
Braun went to Colonel Becker on his own with what little
data he had from the VfR's Raketenflugplatz tests in hand.
Becker was impressed, not only with the information von
Braun presented but with his audacity and confidence as well.
Both men knew a partnership would be beneficial, Becker
getting a weapon and von Braun getting the necessary
funding to build the rockets of his boyhood dreams. So
Becker offered von Braun a deal where the army would
become a benefactor of the VfR. Von Braun was thrilled, but
not everyone at the Raketenflugplatz shared his enthusiasm.
Some wanted nothing to do with the military because of the
restrictions this partnership would force upon their work;
they wanted to build exploratory rockets, not missiles. Other

members were personally wary of forging a partnership with the military lest a conflict break out and force them into the war machine.

The VfR's overall trepidation didn't dampen Becker's interest, and so he made von Braun a second offer: a job developing liquid rockets for the army and a doctorate degree. Colonel Becker was also a professor at the University of Berlin. He could arrange for von Braun's army work reports to be accepted as a thesis, though the material would be confidential. Von Braun accepted and, at just twenty years old, was formally hired by Dornberger and began working for the army on October 1, 1932. The military's poaching of the brightest minds marked the end of amateur rocketry in Germany and the beginning of the rocket's revival as a weapon.

Dornberger recruited other members of the VfR to gradually fill out the ranks of his new liquid-propulsion rocket program, among them Arthur Rudolph and Walter Riedel. With better equipment and facilities at Kummersdorf West, the former VfR engineers stumbled through new rocket designs, trial and error serving as their greatest teachers.

Von Braun's first test for the army came just two months after he was hired. December 21 was a clear, cold night at Kummersdorf West as he, Riedel, Dornberger, and another technician named Heinrich Grunow made their way to an outdoor test stand. Three concrete slabs eighteen feet long and twelve feet high formed an open enclosure that could be closed with a set of folding metal doors. The roof made of wooden slats covered in tar paper was rolled back for the test. Inside the test space that night was the first rocket motor von Braun developed under Dornberger's leadership, a twenty-inch-long pear-shaped engine made of an aluminum alloy called duralumin designed to generate 650 pounds of thrust. Pipes and wires led from the test setup into the control room on the other side of one wall, offering technicians Riedel and Grunow protection from the engine test as they fed fuel and oxidizer into the combustion chamber. Dornberger, meanwhile, took shelter behind a tree.

Once the chamber was filled, von Braun, as the junior member on the team, manually introduced the flame to start the combustion reaction. He carefully picked up a can of flaming gasoline with a twelve-foot-long pole and nudged it into the path of the alcohol and liquid oxygen issuing from the engine's nozzle. In an instant the test stand was engulfed in flames as an explosion ripped the folded doors from their hinges and sent hot metal flying in all directions. The fire died to reveal a mess of charred metal and wires smoldering with thick, noxious smoke. Riedel and Grunow emerged from the control room to find von Braun and Dornberger miraculously unscathed. The culprit turned out to be delayed ignition. In the seconds before von Braun had pushed the gasoline under the engine, enough alcohol and liquid oxygen had built up in the combustion chamber to create an explosive environment. A new lesson was learned, and work continued at Kummersdorf West. A month later the propulsion demonstrations resumed in the rebuilt test stand.

As 1933 wore on, Dornberger's team began building on their early concept rockets to develop their first product specifically for the army. Called the Aggregate-1, or A-1, it marked a significant shift now for the former VfR engineers. At the Raketenflugplatz, the team had designed their rockets with the engine mounted in the rocket's nose with the fuel tanks behind it, not unlike an automobile. For the A-1, they mounted the engine in the rear, below the fuel and oxidizer tanks when the rocket was standing vertically. As rockets became bigger, this simple redesign promised to prevent the rocket's exhaust from incinerating the tanks. Rather than a firecracker, the A-1 resembled a huge artillery shell one foot in diameter and 4.6 feet long. The team also placed an eighty-five-pound flywheel, an instrument that increased the rocket's momentum to make it more stable in flight, in the rocket's nose.

Problems with this new arrangement gradually came to light as testing revealed the design was nose heavy. Its center of gravity was too far from its center of thrust to be stable, even with a gyroscope mounted in the nose. Problems aside, following a series of static fire tests toward the end of 1933, Dornberger's team opted to launch the A-1 anyway. If it failed, as they expected it would, at least they would learn something in the process. The rocket was loaded onto a launch stand, the fuel and oxidizer started flowing, and the ignition lit. In a flash the rocket exploded in a massive fireball, leaving behind a pile of twisted metal. It was the same problem of delayed ignition; an excess amount of pressurized fuel and liquid oxygen had built up in the combustion chamber a fraction of a second before the engines were lit, leading to an explosion.

As 1934 dawned, Dornberger's team took the lessons learned from the disastrous A-1 and set to developing their follow-up rocket, the A-2. This rocket had the same dimensions, shape, and engine as its predecessor, but the placement of its instruments was different. In the A-2, the gyroscope was placed in the center of the rocket's body rather than the nose, something the men hoped would solve the stability problems of the A-1. Two identical A-2s named Max and Moritz (after two popular cartoon characters) were built by the end of 1934.

While the group at Kummersdorf was busy bringing the A-2 to life, the Raketenflugplatz died, killed not by an exploding rocket but by a water bill. With unused buildings falling into disrepair, leaking pipes and faucets became a problem. And while the rent of ten marks a year remained affordable, the water bill from the constantly running taps was crippling. The VfR was forced to relinquish the site to the city of Berlin, and without a space to work the society was all but dead. Amateur rocketry couldn't survive, leaving Dornberger's group at Kummersdorf West as the lone functional rocket group in Germany.

The shift in the landscape of rocketry was nothing compared to the changes taking place in Germany. On August 2, 1934, German president Paul von Hindenburg, the final vestige of the Weimar Republic, had died in office of lung cancer. Immediately afterward, the Law Concerning the

Highest State Office of the Reich was made public, rolling the offices of president and chancellor into one, a single office of der Führer (the leader). The effect was that Chancellor Adolf Hitler was pronounced Führer without a formal election. By extension the National Socialist German Workers' Party seized power and control of the country's assets, including its military. Dornberger's group was now, however indirectly, in the hands of the new national party.

But the changing political landscape did nothing to change the day-to-day workings at Kummersdorf. The year 1934 was devoted to Max and Moritz, both of which were launched successfully from a site on Borkum Island in the North Sea a few days before Christmas. The updated engine and new gyroscope configuration were steps toward a properly functioning missile. The reward for this first significant positive result was an influx of army funding for the group to build the next rocket in the series, the A-3. The twin rockets also brought the rocket group to the attention of Germany's meager remaining air force, the Luftwaffe.

That Germany had an air force at all was a poorly kept secret at best. In the early 1930s, Hermann Göring, the minister of aviation, disguised programs to train military pilots as recreational flight training under the League of Air Sports. But by 1935, the ruse was dropped and the Luftwaffe was formally established, an unmistakable military creation of the Nazi regime. And the new Luftwaffe was very interested in the liquid-propulsion engines the army was working on, not for rockets but for rocket-powered airplanes. Major Wolfram von Richthofen, cousin of the famous "Red Baron" Manfred von Richthofen, visited the Kummersdorf team to request that they develop an alcohol and liquid oxygen engine for an airplane, one that would serve as a secondary engine to a traditional propeller propulsion system. Von Richthofen went so far as to request that a team of his own men move into Kummersdorf West to develop a rocket engine that could power an airplane, a request the army's group obliged. Impressed by the Kummersdorf group's early successes, the Luftwaffe was excited by the prospect of engines that could get heavy bombers into the air and turn small fighter planes into formidable flying weapons, easily outstripping enemy aircraft.

As von Richthofen's plans advanced in the summer of 1935, Wernher von Braun found himself one day sitting in the wingless fuselage of a Junkers airplane bolted firmly to a test stand. Mounted behind him was a rocket engine, the same 650-pound thrust engine that had powered the twin A-2 rockets Max and Moritz. Von Braun threw a switch in the cockpit to fire the engine, and a spear of flame shot out behind the plane. The flame erupted with enough force that, even without wings on the fuselage, von Braun would have shot forward into the air. Even this static test exhilarated von Braun. He had just experienced firsthand the power of his rockets as a means for travel, and it had whetted his appetite for more powerful rocket travel.

Between the new contract developing rocket engines for the Luftwaffe and ongoing work on the A-3 to deliver a functional offensive rocket to the army, Dornberger's team was fast outgrowing the facilities at Kummersdorf. The suburban location was also becoming a problem; increased activity made it harder to maintain the necessary military levels of secrecy at a site just seventeen miles outside of Berlin. It was clear they needed a new location, somewhere isolated where they could set up larger test stands and a firing range to safely launch increasingly large missiles. It also made sense to merge the army and the Luftwaffe's rocket programs into one site to avoid unnecessary duplication.

The search for the perfect test site began in the summer of 1935, and the solution came through the von Brauns. As a child, Wernher von Braun had gone on family hunting trips to the northern German island of Usedom, a quiet, secluded island bordered on one side by the Baltic Sea and separated from the mainland by the Peene River. On the northwestern edge of Usedom was a village called Peenemünde, which literally means "mouth of the Peene," and marked the spot where the Peene flows into the Baltic. The isolation made it perfect for a discreet arms development site, and its coastal location was perfect for a safe firing range; rockets could fly safely over unpopulated areas and land harmlessly in the water.

In the spring of 1936, the German Air Ministry negotiated the sale of a tract of land 2.5 by 7.5 miles near Peenemünde for 750,000 marks, a paltry sum compared to the funding set aside to develop the site. Already cleared to build a new facility for the army, Colonel Becker teamed up with Dornberger, von Braun, and von Richthofen to present their idea of using Peenemünde to General Albert Kesselring, the Luftwaffe's chief of aircraft construction. They quickly sold him on the plan for a joint facility, which led to discussions of funding. Von Richthofen immediately put up five million marks for the Luftwaffe's contribution. Not to be outdone, Becker put in six million marks for the army's allowance for the new rocket facility. Construction began at Peenemünde almost immediately, and while it marked the beginning of a collaborative effort between the army and the Luftwaffe, the latter made another acquisition at the same time that spoke to its separate plans for rocket engine development.

Though published and working in Germany, Hermann Oberth's influence had spread beyond his adopted homeland. Like Valier and von Braun, Austrian-born engineer Eugen Sänger had been inspired by the possibilities of using rockets for spaceflight after reading Hermann Oberth's works on the subject. And like his inspiration, Sänger's own doctoral dissertation on rocket propulsion at the Technical University in Vienna was rejected in 1931 on the grounds that it was too fantastic to be plausible. Also like Oberth, rejection hadn't pushed Sänger out of academia or away from rocketry. By the mid-1930s, he had secured a position as a professor at his alma mater and had a preliminary spacecraft design under development. Sänger imagined a future where commercial flights would carry passengers and cargo through the upper stratosphere to reach any point on the globe within an hour of launching. And so he designed a system he believed might bring this future to pass.

At the heart of his system was a vehicle around ninety feet long with a wingspan measuring close to fifty feet. Roughly cylindrical, the vehicle tapered to a point at the front end and featured small wings and a flat underside, which was the crux of his design. Sänger theorized that if the vehicle could reach a high enough speed and altitude, it could be made to bounce off the atmosphere as it descended from its peak height, much like a stone skips across a calm pond. By harnessing the energy and gliding along an oscillating flight path, he thought, the vehicle could cover a significant distance before making an unpowered landing on a runway, landing like a traditional airplane.

Sänger needed a rocket to get his vehicle up to high enough speeds and altitudes to begin the gliding descent. The launch system he imagined had the glider mounted on a rocket-powered sled. The sled would travel along an inclined monorail track for a horizontal rather than vertical launch, making for an easier flight for passengers and ensuring a simple ballistic flight path. Once airborne, the vehicle's own rocket engine would fire to get it up to altitude, burning every bit of available fuel along the way. The fuel expended, momentum would carry the glider higher until gravity took over and began the vehicle's descent. It would bounce and glide until reaching the lower atmosphere. At that point, it could fly like a traditional glider all the way to the runway. But this was only the first version. With enough rocket propulsion, Sänger knew, that same glider could fly fast enough to leave the Earth's atmosphere altogether and go into orbit.

Sänger was certain that the first iteration of his vehicle could be built and flown with existing technology. All that was missing was the necessary propulsion system, the constant pressure combustion chambers and tanks, to propel the vehicle from its starting point on the sled then boost it into the upper atmosphere. That was something Sänger couldn't develop without laboratory space and funding. And so, like von Braun had done while at the Raketenflugplatz, Sänger turned to the military for sponsorship.

Securing military funding for his vehicle meant Sänger had to pitch his aerospace transportation system as a weapon. So he added bombs to the vehicle, turning it into a manned long-range bomber, and because his glider could theoretically reach any point on the planet within an hour, he called this weaponized version the antipodal bomber.

He envisioned two basic mission types for his antipodal bomber. The first was a point attack, a precision technique that would have the pilot release his bomb from a moderate altitude while flying at a moderate speed. Timed right, he could destroy a bridge, a building, or a tunnel entrance with a single device. The second method was an area bombing attack, which traded precision for might—damaging a larger swath of land, such as a city. This type of attack had the pilot drop his bomb from altitudes as high as a hundred miles while flying at much faster speeds. In both cases, the pilot could continue his gliding flight to a safe landing on a runway, burning any remaining fuel to extend his flight as needed. In the rare event that a target site was too far from a safe landing site, the pilot could ditch the glider, parachuting to safety himself while letting the glider crash.

After a brief stint with the Austrian Nazi Party and its Schutzstaffel—the "Protective Echelon" known as the SS—in 1933, Sänger took his antipodal bomber idea to his native country's army. But the Austrian military wasn't interested; the Austrian National Defense Ministry said it couldn't seriously consider the skip-glide system. Sänger designed it with a liquid oxygen and hydrocarbon combustion system, a poorly understood chemical reaction with a high risk of explosions. The Austrian Army thought it unlikely that such a volatile combination would ever become a practical means of propulsion. Undaunted, Sänger continued exploring liquid rocket propulsion on his own while also pursuing his job as an engineer for a Viennese construction company. In 1934, he took his antipodal bomber to the German Army, but again it failed to secure sponsorship. Because Sänger wasn't German-born, a requested security check into his background submitted to the Sturmabteilung, the "Assault Division," or SA, went unanswered. It didn't help that his work was hardly more advanced than the army's own. Rocket expert von Braun advised the German Air Ministry not to hire Sänger. The army ultimately passed on Sänger's potentially duplicative proposal but advised him to take the antipodal bomber to the Luftwaffe.

The Luftwaffe had no problem with Sänger's Austrian background or with any similarities his technology bore to von Braun's. The engineer was recruited to become a member of the Research Division of the Technical Office of the Göring Institute, joining the aeronautical research laboratory it was planning to build near Braunschweig in north-central Germany. The appointment also granted Sänger funds to establish a separate rocket research facility some distance away at Trauen, which was given the cover name of Aircraft Test Center to hide its existence as much from the army as from the public. The lab at Trauen was also a reaction to the new site at Peenemünde that promised to give von Braun a hefty increase in space and resources to bring his bigger A-series rockets to life. Hermann Göring, minister of aviation and commander of the Luftwaffe, wasn't about to be outdone by Becker, Dornberger, and the army group. He sunk as much as eight million reichsmarks into Sänger's winged bomber project.

By this point, Dornberger's group at Kummersdorf West had several years' lead over Sänger, and the Army's A-3 was far closer to being flight ready than any portion of the antipodal bombing system. The A-3 had some significant advances over its predecessors, notably the addition of an active three-axis gyroscope assembly for control in place of the earlier single gyroscope. The new system meant the rocket could launch without a support structure or guiding rails, making it the first to begin its flights from a freestanding position. The A-3 also featured a liquid nitrogen pressurization system equipped with a heater built into its liquid oxygen tank. As the liquid nitrogen was heated, it boiled off, forcing propellants into the combustion chamber for a more efficient and powerful burn. The A-3 was also the first rocket designed to reach supersonic velocity, flying faster than the speed of sound during its powered ascent while carrying a heavier payload to its target. These changes made the A-3 the team's largest rocket. Gone were the days of transporting a rocket on top of a car. This new rocket could only be transported to the launchpad by rails.

By the spring of 1937, simple but functional residential quarters were built on the southern portion of the site at Peenemünde, allowing the first wave of staff to relocate to the island facility. Workstations and test stands were meanwhile under construction on the island's northern edge, closer to the Baltic. Dornberger and von Braun were among the first to arrive, and to the delight of the men who were slowly taking up residency, the construction crews had left a large portion of the local vegetation intact. It wasn't a stylistic decision. Rather, the trees offered natural camouflage for the materials, buildings, and the rail lines that would soon be transporting rockets to their launchpads. The new facilities also gave the rocket team space to grow. Von Braun, newly promoted to technical director of the group at the time of his move to Peenemünde, brought friends and former VfR colleagues, including Walter Riedel, Arthur Rudolph, and Klaus Riedel, to fill out his team.

Though von Braun's promotion gave him a greater responsibility over the rocket program, it also came with the caveat that he join the Nazi Party. From his beginnings with the VfR to his position working for the army at Kummersdorf West, he had always been a civilian. Now, if he wanted to retain his position as a leader of German rocketry, he would have to pledge his allegiance to the Reich. Von Braun did have the option of not joining the Nazis and forsaking the half decade he had spent developing the Aggregate series, but it was not an appealing alternative. He joined the Nazi Party on May 1, 1937.

With the facilities at Peenemünde still under construction, work on the A-3 continued at Kummersdorf West, and the model grew into an absolute monster. Standing 21.3 feet tall and measuring 2.3 feet around at the widest point, the rocket's long body tapered to a point at the nose, while the base featured small stabilizing fins. Fueled and ready for launch, the A-3 weighed 1,650 pounds, dwarfing the rockets that had come before it. Far too big to launch near populated areas, the Kummersdorf group set up a temporary test stand for the rockets on the sandy island of Greifswalder Oie, just off the northern tip of Peenemünde. They poured concrete, dredged a harbor, and lay rails to transport the rockets. By the beginning of December, they were ready to launch their newest creation.

With a boatload of dignitaries watching from a floating vantage point on the Baltic, the first A-3's launch was problematic from the start. The rocket had been painted with water-soluble green dye that would act as a marker when it landed, but as condensation from the chilled liquid oxygen built up on the rocket's body while the tanks were filled, the paint ran down the length of the fuselage until it reached the electrical cables connected at its base. The mix of condensation and paint shorted the cables, delaying the launch. The electrical issue was eventually resolved, and the rocket took flight over the Baltic. His eyes trained on the rocket, Dornberger watched in shock as the A-3 turned the wrong way around its longitudinal axis running from nose to engine and started flying directly into the wind. It started tumbling as the engine burned through all its fuel, and the instant the parachute deployed it was caught in the wind and forced right into the rocket's hot residual exhaust and started to burn. Without a parachute, the rocket had no recovery system. Dornberger watched, helpless, as the rocket tumbled into the Baltic. Fog rolled over Greifswalder Oie, putting a hold on further launches and giving Dornberger's team time to figure out what had gone wrong.

Days later, von Braun and Dornberger determined that the parachute was the most immediate cause of the tumbling motion. It was also the easiest problem to solve—the parachute could simply be removed, and the rocket would be fine to splash down into the Baltic on its own. They readied another A-3 for launch and, when the fog finally cleared from

Greifswalder Oie, they moved it out to the sandy island for launch. This second flight saw the rocket rise beautifully from the launchpad before tipping over and tumbling into the Baltic. The next launch saw the same troubled flight path repeated. Von Braun and Dornberger were forced to concede that for all its advances and sophistication, the A-3 was a dud. They figured the guidance system was the root cause of its problems, one that merited another series of test rockets. Because the A-4 was already earmarked as the first production rocket for the army, they skipped to the A-5 as the next proof of concept test rocket. They had to figure out the guidance problem if the A-4 was going to reach its target every time.

Von Braun and his team of rocketeers set to work solving the A-3's guidance problem while Dornberger was forced to contend with a larger, bureaucratic issue. Unforeseen development costs had pushed the Luftwaffe out of Peenemünde, leaving the site an army-only facility that couldn't afford to foot the bill alone. What Dornberger needed now more than a rocket was a patron rich and powerful enough to sponsor his team's continued work. But there were far larger problems brewing on a national scale that diverted attention from the esoteric rocket team's problems. Hitler's Nazi Party had started flexing its muscles and making its true might known throughout Europe. In March of 1938, the independent Austrian government fell and the German Army moved in swiftly to take its place. On the night of November 7, a seventeen-year-old Jewish teenager killed a counselor at the German Embassy in Paris. This became the rationale behind an organized wave of pogroms aimed at Jews throughout Germany and German-occupied lands two days later. Gangs of Nazis smashed Jewish-owned shop windows. The Night of Broken Glass, Kristallnacht, ended the following morning with some thirty thousand Jews rounded up and carted out to the first concentration camps.

Against this increasingly hostile political backdrop, Dornberger found a solution to his sponsorship problem and von Braun got a strange twenty-seventh birthday present. On March 23, 1939, Adolf Hitler arrived at Kummersdorf West. The Führer had been Germany's leader for six years, Dornberger had been in charge of the country's rocket program for nine years, and finally the two men met to discuss rocketry's role in Germany's future. Dornberger, along with von Braun, gave Hitler and his small entourage a tour of Kummersdorf West, showing off their test stands and pieces of their hardware. They used a cutaway model of the A-3 to show Hitler what sophisticated instrumentation they had developed, and illustrated the power of their current A-5 with a rocket stripped down to show all its inner workings. They performed a static fire test of an engine to show its power in action. But throughout the day, the Führer was uncharacteristically silent. Typically fascinated by armaments and eager to ask probing questions about specific pieces of technology, Hitler seemed disinterested in the rockets he was being shown. His eyes, thought Dornberger, seemed unfocused, as though his mind was miles away. Even the roaring rocket engine test failed to garner any reaction from the Führer. Not until the men all sat down to lunch together at the end of the tour did Hitler ask any questions. Over a vegetarian meal, the Führer asked how soon the A-4 would be ready and how far it would be able to travel.

An outwardly unimpressed Hitler left a disappointed von Braun and Dornberger in his wake as he departed Kummersdorf West. Without additional money, their rockets wouldn't come to life, and without interest from the Führer, they couldn't expect any increase in funding. The pair couldn't help but wonder if they had demonstrated a small rocket launch whether the day might have ended differently, whether seeing a launch would have convinced Hitler the rockets were worth funding. The cutaways and static tests had perhaps done a poor job of illustrating just what kind of power the army's rockets were capable of to someone familiar with the weaponry men wielded in trenches during the First World War. The potential of rocketry remained untapped, and the program was left with no priority status in a country that seemed increasingly on the verge of another war.

The threat of war became real on September 1, 1939, when Germany declared war on Poland. The aggression, ostensibly due to a hostile attack from Poland, was in fact an attack faked by the Nazi government. A month after their country went to war, Dornberger and von Braun had an A-5 ready to launch. Like the A-3, the A-5 stood 21.3 feet tall and measured 2.6 feet in diameter. Tapered to a point at the nose, it housed a thirty-three-hundred-pound thrust engine in its thick base. All told, this rocket was slightly heavier than its predecessor, weighing in at two thousand pounds, but it addressed the problems that had dogged the A-3. On its first launch, the A-5 rose off its launchpad and followed a perfectly straight trajectory before disappearing into the clouds. At the moment the onboard fuel was used up, von Braun released the parachutes by radio command and watched, alongside Dornberger, as the rocket reappeared from the clouds, suspended from its parachute perfectly on time. It fell softly into the Baltic where it bobbed patiently in the water. It was in perfect condition. Had it not been completely waterlogged, it could have been launched again.

In spite of the A-5's success, Hitler remained unmoved. These missiles were still in a relatively early stage of development, and the Führer only wanted weapons that could be used immediately. He wasn't interested in weapons that would be available months or years down the line. In February 1940, Hitler ordered all weapons development programs that weren't ready for combat struck from the priority list. This meant the rocket program at Peenemünde would continue limping along with meager funding.

Hitler might have remained cold, but there were other factions within the Nazi regime that were quite interested in the rockets coming out of Peenemünde—namely the SS, the muscle behind the Nazi regime that functioned on the strength of volunteers who believed in the Nazi cause. Dornberger knew the SS was interested in somehow getting a foothold in the rocket team's work, but it turned out it was also interested in recruiting Wernher von Braun to join its ranks. On May 1, 1940, SS Colonel Müller arrived in von Braun's office. He was there on behalf of Heinrich Himmler, the strict schoolmarmish head of the SS who was known for his penchant for punishment. Müller had orders for von Braun to join the SS, he told the engineer. Von Braun rebuffed the request, explaining that he was too busy with his rocket work. Müller countered this thinly veiled refusal with the assurance that any time commitments to the SS would be minimal, and von Braun could join right away as an Untersturmführer, a lieutenant. Von Braun again held Müller at bay, this time asking for a few days to consider the invitation, a request the SS representative honored.

Fully aware of the potential political implications if he became involved with the SS, von Braun immediately took the matter to Dornberger, who was still his military superior. Dornberger's opinion confirmed von Braun's feelings; he told his younger colleague that joining the SS was the only way he could continue working on the A-4 program with the army. If he agreed, the SS, and particularly Himmler, would have its coveted direct link to the program at Peenemünde. The alternative was to forfeit his work and likely be sent to perish in a work camp. Again, a desire to continue his rocket work weighed heavy, and von Braun wrote to Müller with his decision. Within two weeks he received a reply saying his application to join the SS had been approved by Himmler.

His new affiliation didn't have an immediate impact on von Braun or his work at Peenemünde, which continued to want for financial support. Business continued as usual. Von Braun visited Sänger's laboratory at Trauen and returned the favor by bringing the Austrian engineer to visit the site on the Baltic. The two rocket scientists crossed paths again and shared research at a hypersonics conference at Peenemünde in October 1940, continuing their independent programs. For von Braun, his life remained devoted to getting the A-4 off the ground while the war intensified around him.

1940 saw the Nazis conquer Poland before attacking Belgium, the Netherlands, and Denmark, and occupying a large portion of France. Hermann Göring had the Luftwaffe bomb England, the country Hitler needed to conquer if he was going to control all of Europe. In the summer of 1941, Hitler pushed the German front into Russia, advancing over five hundred thousand square miles of land. It seemed like only a matter of time before the war would spread to the United States.

U.S. Army Chief of Staff General George C. Marshall had been watching as the clouds of war gathered over Europe, and he worried that his American forces wouldn't be ready should the country be forced to enter the conflict. With less than 190,000 active personnel, a portion of which were stationed overseas, the United States Army existed almost as a token establishment in the country and hadn't been through any kind of field exercise since the First World War. Worried about America's overall unpreparedness, Marshall urged President Franklin D. Roosevelt to initiate the Protective Mobilization Plan—a draft—which went into effect one week after Hitler's attack on Poland. As the American Army grew, new conscripts went through small-scale training exercises, and by June 1941 the U.S. Army comprised more than 1.4 million men. But Marshall's concerns remained. The men needed a large-scale training exercise to ready them for battle, and their commanding officers needed the same training to learn how to handle forces in a combat situation. Field maneuvers had long been a staple in army training, and the war's development in Europe afforded Marshall the rare opportunity to conduct a massive-scale field maneuver training exercise before formally entering the war. It was a way for mistakes to be made at home, not overseas where men's lives were on the line.

In Louisiana on the morning of September 15, 1941, dirt roads turned to mud as rain soaked fields and the nearly five hundred thousand soldiers who had begun battling for control of the Mississippi River. The Red Army of Kotmk, a fake country made up of Kansas, Oklahoma, Texas, Missouri, and

Kentucky, was commanded by Lieutenant General Ben Lear. It was fighting the Blue Army, the army of the equally fictitious country of Almat comprised of Arkansas, Louisiana, Mississippi, Alabama, and Tennessee, led by Lieutenant General Walter Krueger. In his bid to win the war game, Krueger assembled a staff of men he considered brilliant if little-known, among them Lieutenant Colonel Dwight Eisenhower, who served as his chief of staff.

Eisenhower's previous military experience had been largely focused on strategy. He hadn't seen combat in the First World War; much to his chagrin his repeated requests for overseas duty had been denied. Instead, he had spent the war commanding a unit that trained tank crews, and while his unit never saw combat he did gain a keen sense of military strategy and organization. After the war he spent time reading through documents to understand exactly how the U.S. Army had moved through France, an academic study of war he now recalled during the Louisiana Maneuvers. He skillfully directed his men, outmaneuvering the opposing forces, and when the exercise ended three weeks later the Blue Army emerged victorious. The coup earned Eisenhower a place in Marshall's little black book where he kept a list of all the officers he believed could lead the U.S. Army in battle in Europe. Journalists covering the Maneuvers were equally impressed with Eisenhower; the smart, handsome young officer graced a number of front pages when the training exercise was over.

Two months later and newly promoted to the rank of general, Eisenhower spent the morning of December 7 catching up on paperwork. He took a brief break to lunch with his wife, Mamie, before retiring for a short Sunday afternoon nap. He'd barely fallen asleep when the phone rang. It was Ernest R. "Tex" Lee, Eisenhower's aide during the Louisiana Maneuvers. The Japanese had attacked Pearl Harbor, Lee said, and America's Pacific Fleet had been destroyed. Within hours, orders started pouring in for Eisenhower's Third Army, which had served as the Blue Army of the Louisiana Maneuvers, from the War Department.

Four days later, Germany and Italy declared war on the United States, and the phone on Eisenhower's desk rang again. This time General Walter Bedell Smith was on the line with orders from General Marshall for Eisenhower to get on a plane to Washington immediately. He had been called to join the War Plans Division of the army general staff. The United States had joined the war with General Marshall commanding the U.S. Army's entrance into the arena. Eisenhower was brought in as commander of the European Theater of Operations, a role that would demand the abstract understanding of a large-scale war effort he had demonstrated during the Louisiana Maneuvers.

The war gained steam with its newest participant arriving in Europe, and still Germany was without a field-ready missile. But the Peenemünde team was getting closer. Just before noon on the morning of June 13, 1942, a forty-seven-foot-tall rocket 5.5 feet in diameter rose off its launchpad, somewhat unsteadily at first then gaining confidence as it climbed higher into the sky. Von Braun was elated as he watched a decade of work take flight for the first time. Albert Speer, Hitler's chief architect turned minister of Armaments and War Production, looked on in astonishment as the rocket disappeared into the clouds over Peenemünde. The other technicians and visiting military personnel on hand cheered as the rocket rose, though they could scarcely be heard over the thundering engine. The A-4 could fly, and it was a powerful rocket. As the gathered crowd celebrated, the sound of the distant rocket changed, growing louder as the rocket veered off course. It exploded on impact, unsettlingly close to where the visiting dignitaries stood watching. The next A-4 took flight on August 16. Again, it rose boldly off the pad before being struck down by failure. A flaw with the guidance system left it carving a jagged supersonic path through the sky before it exploded. But von Braun welcomed failures. Every failed launch brought a new problem to light, and identifiable problems were easier to fix than unidentifiable ones.

The next test came on October 3. Dornberger stood, binoculars in hand, on the flat roof of the Measurement

House at Peenemünde. The sky overhead was clear and cloudless as it stretched over northern Germany. On a television screen was the picture of a rocket painted black and white and gleaming in the midday sun. It was alone, service platforms and personnel having cleared the area. With one minute to launch, Dornberger could feel the tension in the air rise. Looking through his binoculars he could see clouds of smoke issuing from the bottom of the rocket, followed by a shower of sparks that was quickly replaced by a flame. Pieces of wood and grass flew through the air as cables detached from the rocket, which began to climb. Dornberger followed along as it traveled north to rise out of the forest into the clear sky, its thunderous roar reaching him a full five seconds later. He watched as the missile began its scheduled roll and turned almost imperceptibly to the east, all the while traveling faster until it reached the speed of sound. A thin white trail began streaming out behind the rocket, sending the men watching the launch into a panic. The missile had exploded, some yelled, while others were sure the trail was just the liquid oxygen venting. But the rocket kept flying, and Dornberger wept with joy, and when the rocket landed in the Baltic almost five minutes after launching, he could rightfully call the test a success. It was the culmination of a decade of work, and the spaceship, Dornberger knew, had been born with that launch.

Just before the end of 1942, Heinrich Himmler, the ruthless head of the SS, chief of the Gestapo, and the man behind the infamous concentration camps, arrived at Peenemünde unannounced. Dornberger gave Himmler a tour of the site that involved more conversations in the officers' mess than demonstrations; there were no launches planned for that day. It turned out that, in spite of the Führer's apparent disinterest in rockets, there was great interest in the rocket among Hitler's inner circle, and Himmler was there to learn as much as he could about the weapon with the goal of using his position to help Dornberger's team get the funding and the support they needed. What was more, continued Himmler, the rocket's development was no longer just the army's concern. It was now the concern of the German people and had to be protected as such. Himmler wanted to establish his own men on the base. Dornberger was pleased about the interest in his program but balked at the idea of Himmler taking control of the site and suggested instead that the SS take control of security in the surrounding town of Peenemünde. Himmler agreed, though as he left the rocket site, he promised to return.

Himmler did return to the rocket site on June 28, 1943. The visit started with a modest dinner in the officers' mess with a small group that included Dornberger, von Braun, and a handful of local dignitaries and colleagues. After the meal, the group retired to the Hearth Room, a warm space with wood-paneled walls and brass details, for social conversation. Talk eventually turned to rockets, a subject von Braun happily took the lead on. He shared with Himmler his personal history, from his work at Kummersdorf West and Peenemünde to his dream of using rockets to explore space. However, for the head of the SS, rockets were a way to deliver a bomb onto a target from a great distance; they had no use off the Earth. But none of the rocket men in the Hearth Room that night paid much attention to the disparate views on their creation.

Rockets took center stage again the following day. At a quarter past nine in the morning, an A-4 lifted off its launchpad and promptly exploded. A second rocket launched that afternoon fared much better, flying perfectly straight until it disappeared into the clouds. Lost to sight, the continued, thunderous roar of the engine signaled that it was still traveling toward its target. The demonstration gave Himmler a sense of what had the rocketeers so excited, and he left impressed by the group, a favorable opinion that reached Hitler quickly. Less than two weeks later, the Fürher invited Dornberger and von Braun into his inner circle to present on their program at the army's guesthouse in East Prussia.

When von Braun and Dornberger arrived for their meeting with Hitler, they were ushered into a restricted area and left to wait in a room with a projector. They would wait there until five in the afternoon, they were told, at which time Hitler would arrive, and they would promptly begin their presentation. But five o'clock came and went, and still they waited. The time seemed to drag on interminably when the doors to the room banged open, announcing Hitler's arrival. Von Braun and Dornberger sprang to their feet, greeting their host and his entourage. The men all took seats, and the presentation began, von Braun giving a run-through about the mobile launch system at Peenemünde before showing a film of the October 3 flight. Dornberger followed, presenting the hard facts and walking through the development schedule and technical details of the A-4. He showed pictures of the impacts his rockets had left after successful tests. When Dornberger finished speaking, Hitler got up and shook both men's hands.

The Führer was impressed, far more than he had been after his 1939 visit to Kummersdorf West. This time he had immediate questions. Could the payload capability of the A-4 be increased to ten tons, he wanted to know, and how many rockets could Peenemünde produce each month. It was clear that having seen the power of the rocket, Hitler finally believed in the project. He granted the program the priority status Dornberger had coveted for so long, but Hitler also asked for two thousand A-4s each month, a figure well beyond Peenemünde's means. To compensate, the Führer said, there would be new factories built and a new workforce applied to the rocket program. Hitler saw in the A-4 the means to annihilate major cities, forcing his enemies into submission. It could be the secret weapon he needed to win the war. He could only hope it would remain secret long enough to be fully operational.

CHAPTER THREE

The Turning Tide of War

Just before midnight local time on August 17, 1943, eight British Royal Air Force Mosquito bombers edged into the skies over Berlin. As part of Operation Whitebait, each plane dropped its three five-hundred-pound bombs and chaff—masses of radar reflective metal foil, used to mask their true location from enemy tracking stations on the ground. The Germans knew this small fleet was a standard forerunner of a heavy raid and reacted accordingly. Air raid sirens started blaring, sending civilians running for cover as searchlights started sweeping the sky to illuminate the coming battle. The Luftwaffe deployed 150 fighters. The Mosquitos left the area, and for the first time, the Germans were pleased to have aircraft ready and waiting for the coming Allied bombardment fleet.

More than a hundred miles away at Peenemünde, the sound of the air raid sirens woke Wernher von Braun. It was a familiar sound, as were the sounds of planes overhead; aircraft frequently passed over the rocket station on their way to Berlin. A radio announcement confirmed the activity over the German capital, a safe distance away. But then a red marker appeared in the sky over Peenemünde followed by sixteen more, dotting the night with white lights. Von Braun saw them floating in the moonlight, hanging like Christmas trees. Then, fifteen minutes after midnight, a wave of bombers appeared in the skies over the rocket site, releasing their ordnance over the Siedlung, one of the housing estates. Von Braun heard the roar of the bombers followed by the thumping of German antiaircraft guns and immediately ran to take cover in one of the concrete blockhouses.

The sound of gunfire woke Walter Dornberger. He threw on clothes and, remembering that he'd sent his boots away for polishing, slid on his bedroom slippers before running for the air raid shelter. As bombs continued falling from the sky, searchlights illuminating the action from below revealed that German air cover was scant. The Luftwaffe had taken the bait over Berlin and left the rocket site unprotected.

Determined to save his work from the raid, von Braun left the safety of the shelter and ran into the burning building that housed his office. His secretary and a handful of others followed him. Groping along the walls as the building collapsed around them, they made their way into von Braun's office. His secretary carried armfuls of documents from the burning building while her boss threw more materials out of the window to save them from the fire. Dornberger, meanwhile, rushed into the Measurement House to save the team's guidance, control, and telemetry work. All the while, bombers flew overhead, and bullets streaked through the sky. When the Luftwaffe finally arrived, the new planes only added aerial fighting to the chaos.

An hour after the first RAF forces arrived at Peenemünde, the last wave dropped their bombs and turned back toward England with forty fewer planes in their rank. And while Peenemünde had taken a heavy beating, the damage could have been far worse. The sky over the Baltic that night was spotted with clouds, and the radar on the lead planes wasn't working. The bombardiers tried to calculate the correct drop points for their first target markers but missed. Some flares were dropped as markers over the water while others fell two miles south of where they were intended to drop. These first flares had marked the Trassenheide labor camp at Peenemünde and not the leaders' housing building, which had been the Allies' main target. The RAF eventually corrected the error and reached the main buildings on the rocket facility, but that initial miscalculation led to a slight delay that gave the rocket scientists time to take shelter. In all, eighteen hundred tons of Allied bombs rained down on Peenemünde as part of Operation Hydra, killing 180 Germans and around six hundred foreigners in the workers' camp where there were no concrete shelters. Though the facilities had suffered serious damage, the raid failed in its main goal. It didn't take out the leading talent behind the A-4 program. Von Braun, Dornberger, and the core men whose knowledge was vital to seeing the program to completion had survived. They buried their dead colleagues two days later.

The raid on Peenemünde had been in the works for months. The first indication that the Germans were developing a long-range bombardment weapon had reached the Allies in the form of anonymous letters delivered to a British attaché in Oslo. The so-called Oslo report mentioned a site on the Baltic called Peenemünde where tests of long-range weapons were underway, though the Allies had no way to corroborate the claims made in these letters. But evidence appeared before long. Danish fishermen reported seeing objects streaking through the sky leaving flaming trails. Word from the Polish underground was that forced laborers on the island of Usedom had found missile-looking objects in a shed. German prisoners of war told their Allied captors about large rockets they had seen under development. It was enough evidence for the RAF to investigate by aerial photography. In June, the Allies saw unmistakable evidence of finned, torpedolike objects. It was clear these were long-range missiles. The air raid was an attempt to neutralize the threat.

But Operation Hydra had the reverse effect. Not only did bombers fail to kill key German scientists, the raid spooked Hitler into taking increased measures to guard the team at Peenemünde and their valuable work. To protect the newly prioritized rocket program from another attack, the Führer ordered all A-4 production moved underground. To ensure nobody building the rockets could leak secrets, he decreed that the A-4 program use concentration camp labor. On August 20, Hitler appointed Himmler as his new minister of the Interior, which put him in charge of organizing the A-4 program's new subterranean factories. Himmler in turn brought in Hans Kammler, a high-ranking officer. The SS was slowly but surely wresting control of the rocket program, exactly what Dornberger and von Braun had been wary of for years.

The A-4 program began preparations to move from its coastal site to central Germany, a spot southwest of Berlin nestled between the Elbe and Weser Rivers. Its new home was a disused sodium sulfate mine and former motor fuel storage site that was little more than a cavern carved into the Harz Mountains. A nondescript name masked the existence of the future rocket factory: Central Works Limited, or Mittelwerk in German. On August 23, the first trainload of prisoners arrived from the Buchenwald concentration camp. Working twelve-hour shifts, the slave labor widened the tunnel and extended it to the other side of the mountain two miles away. As more trainloads of prisoners arrived, they joined the work cycle. Bunks were eventually built in the cave, which remained poorly ventilated and without a sanitation system. Prisoners were forced to drink what water they could find pooled on the floors. Disease ran rampant in this subterranean hell, eventually called Dora but eerily reminiscent of Fritz Lang's *Metropolis*. When Dora became overrun by prisoners, a second camp called Nordhausen was established to also feed laborers into the rocket factory. As far as the workers knew, the equipment they were slowly bringing into Mittelwerk was from a place called Peenemünde, and it was intended to build a secret weapon for the SS under Kammler's direction.

Four A-4s were built at Mittelwerk before the end of 1943 as proof that the factory was up and running, but they were far inferior compared to the rockets built at Peenemünde. Forced labor by unskilled workers produced a flawed product, and all were sent back for repairs and adjustments at the start of 1944. Still, Hitler remained unwavering in his decision to use prisoners to build his miracle weapon and praised Kammler's efforts to fast-track its production by promoting him to SS Gruppenführer, lieutenant general.

Quality control remained an issue as more A-4s came out of Mittelwerk. The stunning successes of the Peenemünde-built rockets weren't replicated with the Mittelwerk-built versions; fewer than 20 percent of the rockets coming out of this subterranean factory successfully launched and reached their targets. The rushed mass production schedule exacerbated lingering problems in the still-nascent science of rocketry, but so too did sabotage. As laborers figured out what they were building, they did what they could to render individual pieces inoperable, loosening connections or urinating on electric units. The A-4 was the most technically sophisticated weapon to that point, and neither von Braun nor Dornberger was surprised to find it wrought with problems when built in squalid conditions by men with no vested interest in its success. To the engineers, Kammler's plan to fast-track the A-4 risked killing it instead.

In spite of mounting difficulties, Hitler's interest in the A-4 deepened, becoming almost fanatical. So too did Himmler's desire to wrest control of the program from Dornberger. Toward the end of February, he called von Braun into his office, an invitation that did not extend to Dornberger, who remained von Braun's military superior officer. The A-4, Himmler said, was no longer a toy. It was a weapon all of Germany was eagerly awaiting. Wouldn't von Braun like to develop his rocket free from all the army's regulations and red tape? Himmler once again offered von Braun the chance to join his staff, enticing him with the promise of a simplified chain of command with his superiors afforded him by his rank. Von Braun, who had been promoted through the ranks of the SS yearly, was now a Sturmbannführer, a major in the SS. But von Braun had had enough of Himmler's meddling. Loyal to Dornberger and the program he'd been a part of for more than a decade, the engineer refused Himmler's offer.

A month later, days before his thirty-second birthday, von Braun was roused by a persistent knocking in the early hours of the morning. It was an unwelcome interruption. He'd returned to his room at the Inselhof Hotel near Peenemünde hours before from a business trip, and was annoyed to be woken so soon after getting to bed. He grudgingly rose, his tired irritation turning to shock when he saw officers of the Gestapo at his door. Shock soon turned to anger; these men should know better than to wake one of Germany's scientific and intellectual elite at such an ungodly hour. But von Braun fell silent when one of the officers started reading official orders requesting he accompany them to police headquarters in nearby Stettin. He wasn't under arrest, one of the officers assured von Braun, they were simply putting him in protective custody. Von Braun consented and left with the officers, joining them as they paid visits to three of his colleagues, his brother Magnus, Klaus Riedel, and Helmut Gröttrup. Once in Stettin, all four were promptly placed in individual holding cells on the top floor with no explanation as to their incarceration. The next day, the guards allowed the men to share a food packet von Braun's chauffeur brought by the jail to mark his birthday, but soon the rocketeers were back in their individual cells.

As the days stretched into two weeks, the men learned they were all being held on charges of treason against the Reich, a crime that usually ended in execution. It was his love of spaceflight and penchant for heavy social drinking that had gotten von Braun into trouble. At a party weeks earlier, he'd mentioned in casual conversation that he foresaw the war ending poorly for Germany and added that all he'd ever wanted to do with his rockets was launch them toward other planets. It was a similarly damning admission to the one he'd voiced to Himmler at Peenemünde a year earlier. Magnus von Braun, Riedel, and Gröttrup were being held on the double charge of voicing similar opinions and being close friends of Wernher's.

Neither his position as a top military scientist nor his membership in the SS could save von Braun now, but Dornberger could. That the SS was unable to find any evidence that the leader of the army's rocket program knew about or shared his subordinates' treasonous tendencies left Dornberger free to defend von Braun and his colleagues. Dornberger knew all four men were indispensable to the A-4 effort, and he suspected that their imprisonment was more than likely another attempt by the SS to strong-arm control of the program away from him and the army while simultaneously lighting a fire under the scientists to get their rockets successfully flying sooner.

By early April, von Braun knew the charges against him would likely bring about a days-long, if not weeks-long, interrogation at the hands of the Gestapo, and Dornberger had not yet managed to reach the right authorities to free his colleagues. Appealing to increasingly high-ranking officials in the Reich, Dornberger finally found success with Albert Speer, the minister of Armaments and War Production for the Third Reich. Orders for a conditional release in hand, Dornberger swept into Stettin on the second day of von Braun's interrogation to free him. They broke into a large bottle of brandy shortly after leaving the jail. Their three coworkers followed days later.

Though his imprisonment ended before he was formally incarcerated, it forced von Braun to recognize a grim truth. The country he loved and the army that had funded his research for more than a decade had the potential to turn on him on a moment's notice, switching from benevolent patron to lethal threat. His own feelings about and goals for his rockets would have to remain firmly separated from his professional goals if he was to survive the war. His imprisonment hadn't changed his conviction that the war had taken a turn and Germany was unlikely to emerge the victor, and he now had to begin thinking about his own well-being.

He began living a dual life. Publicly he was a dutiful soldier of the fatherland, visiting front line units in full SS dress. Privately, he kept his eyes open for opportunities. He would not only have to find a way to get out of Germany alive, he would have to bring with him his rockets, their blueprints, and ideally his team of engineers. Von Braun wanted to move everything to America. He had been fascinated with America since his older brother, Sigismund, had spent a year there studying law and touring in a Model A Ford. America seemed to von Braun the perfect place to build rockets, an American dream that gradually outweighed his feelings of loyalty to his homeland. He also privately thought that with Germany poised to lose a second major war, moving to the United States might put him on the victor's side should another major international conflict erupt.

Just under two years after Dwight Eisenhower arrived in London armed with the absolute powers of a theater commander, von Braun's prediction of the war ending with a vanquished Germany began taking shape for the Allied forces. In 1942, Eisenhower had orchestrated the successful Operation Torch invasion of North Africa that secured Allied control in the Mediterranean. In the fall of 1943, he had facilitated a cease-fire with Italy. The general's next major move was the invasion of Nazi-occupied France. Code-named Operation Overlord, the planned Allied liberation of France in the west in conjunction with a push from the Soviet Union from the east to reclaim its lost territories in Eastern Europe would be the definitive step toward taking Berlin. Who to command Operation Overlord was the question. U.S. president Roosevelt insisted it be led by an American; British prime minister Winston Churchill and Soviet premier Joseph Stalin agreed. It was widely assumed Roosevelt would choose General Marshall, the army chief of staff. He was the logical choice, but Roosevelt had another thought. The existing command structure was working well with General Eisenhower in charge of the European Theater of Operations and Marshall serving the president and the high command in Washington. If Marshall were to command Overlord, Eisenhower would take his place in the United States. Roosevelt couldn't deny the value of Eisenhower's firsthand knowledge of amphibious operations and the conditions in Europe, knowledge that Marshall did not have. On a personal level, the president also liked that Eisenhower displayed none of the posturing so common among high-ranking military officials.

At a summit meeting between Roosevelt, Churchill, and Stalin convened at the Soviet Embassy in Tehran, Iran, on November 28, 1943, the question over who would command Overlord was again on the table. Roosevelt's indecision lasted the full five days of the conference. He knew Marshall deserved the command, but his confidence was ultimately in Eisenhower's abilities; the general had never seen combat firsthand, but Roosevelt felt he had an unparalleled understanding of large-scale operations gained from years of study. The president met with Eisenhower in Tunis and told him to start packing. Eisenhower was going back to London as the commander of Operation Overlord. The news made the general grin like a schoolboy.

Eisenhower spent six months meeting with military and political leaders to solidify plans for Operation Overlord. The invading forces grew to include American, English, and Canadian paratroopers and forces crossing the English Channel to land by sea at the beach in Normandy, France. They would first secure a foothold in France, liberating and rearming the country, then press onward toward Germany. Aerial bombing campaigns would aid the invasion forces, destroying roads and rail lines to stop German forces from advancing toward the Allies. Precision bombings of aircraft production plants, fuel storage depots, and airfields would further weaken the German defense. The ideal date for the invasion was set: June 7 or 9. Those dates marked a confluence between tides, moonlight, daylight hours, and favorable weather for an invasion; the weather had to be just right for all the land and air troops to arrive safely at their targets. Missing that window would push the invasion back until at least June 19, and Eisenhower didn't want to miss the first chance to attack. This was a crusade in which he would accept nothing less than a full victory to end the war in Europe.

As the window for Operation Overlord neared, the weather refused to cooperate. Low clouds brought poor visibility, high winds threatened to complicate paratroopers' landings, and rolling waves promised extreme difficulty in adjusting naval gunfire and the safe navigation of landing craft. At ten o'clock on the night of June 4, Eisenhower was forced to postpone the invasion. Six hours later he met with his commanders against a backdrop of gale-force winds and pelting rain. But meteorologists found a glimmer of hope. The weather responsible for the delay was moving fast, and there would be a window of calm early in the morning of June 6. Eisenhower made the call. Operation Overlord was back on. The night before the planned invasion, he visited troops of the 101st

Airborne Division near Newbury, Wiltshire, in England. They would be the first men to land in France.

Just after one o'clock in the morning local time on the morning of June 6, the first paratroopers began dropping near the beaches at Normandy, seizing roads and bridges. Within half an hour alarms were sounding. The German generals knew they were witnessing the beginnings of an invasion, but they also assumed it was a feint; they were expecting a major invasion at Calais, not Normandy. At four o'clock, aerial bombing began. By dawn, the first troops began storming onto three beachheads.

Too late did the Germans recognize that the forces at Normandy were the main thrust of the invasion, and Hitler refused to believe it at all. Although Allied troops had pushed almost six miles into France, Hitler ordered his generals to hold off on sending counterattacking divisions until he could see what developed. Then he went to bed. When Hitler woke up at three o'clock that afternoon and realized the gravity of the situation, he not only gave permission for German troops to join the fight, he ordered that they take back the beachhead by the end of the day, a wholly impossible order. When the Sun set, the Allies had gained a strong foothold in Continental Europe. With one hundred thousand Allied soldiers beginning the slow, long trek toward Berlin, Hitler's need for a fully functional miracle weapon with the A-4 only increased. It was time for the Führer to take a more drastic measure.

Less than two months after the D-Day invasion at Normandy, Hitler promoted Himmler to head of the Home Army, and because Dornberger's group reported to the Home Army, Dornberger now reported to Himmler. By extension, Himmler had effectively gained control of the A-4 program. On August 8, Himmler appointed Kammler special commissioner for the A-4 program, a level of authority that Dornberger had never had. Kammler had the power to deploy the A-4 as a combat missile. For Dornberger, it felt as though he'd spent a lifetime lovingly crafting a violin only to watch helplessly as an unmusical brute unceremoniously scraped its strings with a block of wood. There was nothing Dornberger could do. The program he and von Braun had built was now in the hands of Himmler, one of Hitler's most universally feared servants.

Dornberger threw himself into perfecting the rocket to appease his new commander, but the program was worse off for the SS's involvement. Kammler wasn't a rocket scientist. The changes he ordered for the A-4 were more of a hindrance than a help, just token changes to exercise his control over a technology he didn't understand. His efforts ultimately added inefficiency to the program. The abysmal conditions at the Mittelwerk factory and nearby Camp Dora also didn't make things any better. Public mass hanging became the norm for prisoners caught trying to sabotage the rockets, a way to both punish troublemakers and deter others from following in their footsteps; the bodies were typically left hanging for at least half a day. By the fall of 1944, the army's A-4 project was firmly under the command of the Nazis and the SS. Von Braun and Dornberger, by extension, became an integral part of the SS program using slave labor to build long-range missiles. They could be held responsible for deaths on both sides of their rockets' flights.

By late August 1944, the Allies had reclaimed a sizable portion of northern France and were moving closer to the German border. As part of his counterattack, Hitler ordered that the rocket bombardment begin as soon as possible. Executing this order fell to Kammler, who deployed two firing units. Group North moved northwest from Cleves toward the Hague in Belgium to get a good shot at London, and Group South moved north from Baumholder to Koblenz to hit Paris. On September 6, Group South positioned an A-4 on its mobile firing table, and a little after ten-thirty in the morning, sparks began spewing from its engine. The sparks turned to flame and then a cloud of smoke as the rocket came to life and rose slowly off the launch platform. The rocket sputtered and fell back down with a thunk, luckily resting upright on the launch platform. A second A-4 launched a little over an hour later ended with the same failure. Premature cutoff of the fuel supply had grounded both rockets.

Two days later, the missile unit traveled to Houffalize in Belgium and loaded another A-4 onto a launch platform. At eight thirty-four in the morning, the first successful long-range missile attack rose into the sky. It traveled 180 miles and caused moderate damage upon impact near Porte d'Italie in France. At six forty-eight that evening, the second A-4 of the day rose up from Group North's launch site just north of the Hague. It carried its small warhead two hundred miles to London where Allied radar had no way of tracking the incoming supersonic missile. The residents of the London suburb of Chiswick who actually heard the explosion were the lucky ones who survived.

The team at Peenemünde learned of the successful A-4 attack from a newspaper headline. VERGELTUNGSWAFFE-2 GEGEN LONDON IM EINSATZ (VENGEANCE WEAPON 2 IN ACTION AGAINST LONDON). The rocket had been christened the V-2 by Nazi Propaganda Minister Joseph Goebbels and touted as the weapon that would win the war for Germany.

Standing in stark contrast to the V-2 rocket's newfound fame was Sänger's antipodal bomber's increased obscurity. Work at Trauen had never matched the pace of developments at Peenemünde, and Sänger's facility had been closed in the summer of 1942 under the guise of staff conflicts and fuel shortages. Around the same time, Göring had withdrawn Luftwaffe support for a rocket program as well. Sänger's team had been working on a one-hundred-ton thrust rocket motor. With their lab closed, they were left working on far less futuristic technologies, propulsive duct motors that used fans mounted inside a shroud and rockets using nitric acid and diesel oil. But Sänger never abandoned his antipodal bomber. He and his partner, mathematician Irene Bredt, cowrote a full proposal in 1944 called "A Rocket Drive for Long Range Bombers" and began quietly circulating it around scientific circles in which Dornberger traveled in the hopes of securing a benefactor.

News of the Pennemünde rockets' success also reached the United States, though the tone was far from celebratory. Even if the V-2 had come to the war late, it was clear that America's technological advances and air supremacy were being challenged. The V-2 didn't have the range to reach the United States from Europe, but to avoid being on the receiving end of rockets launched in some future war, America would have to master the major German technological advances of this war first. General Henry H. "Hap" Arnold, chief of the Army Air Corps, knew this meant learning about the advanced weapons directly from the German scientists who designed and built the V-2. In November, Arnold sent a memorandum to Theodore von Kármán, an émigré from Budapest who, as one of the leading aerodynamics and propulsion experts in the United States, was teaching at the California Institute of Technology and directing its Jet Propulsion Laboratory. The continued security of the United States, Arnold wrote in his letter to von Kármán, will rest in part in developments by the nation's educational and professional scientists. He went on to emphasize the importance of strong weapons in war since the goal of any war is to destroy the enemy's will to resist. Having said all this, Arnold asked von Kármán to lead the new Army Air Force Scientific Advisory Group with the goal of studying the significance of scientific warfare and development of rocketry and guided missiles. What Arnold wanted in the end was expert opinions on how these technologies might benefit the future of the U.S. Army Air Force.

In December, the AAF Science Advisory Group (SAG) was staffed with experts in fields ranging from aircraft design and aerodynamics to fuels and propulsion. As his deputy, von Kármán selected his friend and colleague Hugh Dryden, an aerodynamicist from the National Bureau of Standards. Dryden had been twelve years old when he saw his first airplane, a fifty-horsepower Antoinette with a top speed of forty miles per hour that left him thoroughly unimpressed. The passenger and cargo payloads of airships outstripped the capabilities of winged machines designed for commerce, exploration, and recreation, the preteen wrote in an English composition just days later. It was an acute insight that earned him a failing grade but hinted at the course his career would take. Following a path of excelled academics, Dryden

enrolled at Johns Hopkins University when he was just fourteen and graduated three years later with honors. By twenty,
he'd added a master's degree to his credentials. He spent the
1920s working at the Bureau of Standards on design characteristics that would allow airplanes to fly faster than the speed
of sound, paying special attention to issues of compressibility,
the phenomenon of air building up in front of a fast-moving
airplane.

This work, done at a time when airplanes were barely
flying at half the speed of sound, led Dryden to the National
Advisory Committee for Aeronautics in 1931. The NACA,
the leading national body for aviation research, had been
established as a rider, a minor add-on to a naval appropriations bill in 1915 to help the United States gain air superiority
during and in the wake of the First World War, a role it would
reprise during the Second World War. Not long after the war
in Europe began, Dryden was named the head of the NACA's
fledgling guided missile section under the Office of Scientific
Research and Development. In this capacity he led the team
that developed the Bat guided missile, a U.S. Navy weapon
that combined an existing airframe with a state-of-the-art
thousand-pound bomb. Gyrostabilized with an autopilot
system and steered by a tail elevator powered by small wind-
driven generators, the Bat was America's first reliable precision
bomb that successfully sank a number of enemy submarines
and earned Dryden a Presidential Certificate of Merit.

The call for Dryden to join the SAG came from von
Kármán himself. The Hungarian physicist told Dryden over
the phone that he had agreed to travel to Europe for General
Arnold's project to assess enemy wartime breakthroughs in
flight, explore German laboratories, and interview their
scientists. Von Kármán wanted Dryden to bring his experience with missiles to Europe as well. Dryden accepted the
appointment, and in December started learning everything
he could about the V-2. One of his overseas tasks would be to
interview its creator, Wernher von Braun.

Assembling the scientists that made up the SAG was one
thing. Getting the group into Europe to carry out their

research and interviews was another. The war was still raging. Not only was it unclear how exactly the scientists would manage their assignment, they would need clearance from General Eisenhower just to enter the European Theater. The general, meanwhile, was singularly focused on orchestrating the final push to crush Germany. He didn't want civilians in the area and didn't need the distraction of a group of American scientists with no military training moving through the country. But by the same token, the scientists needed to find their German counterparts before it was too late. Every day increased the chances of them being bombed by the Allies, captured by Russian soldiers, or even killed by their own German garrison.

Fear began seeping through Germany as 1945 dawned. The East Prussian Offensive had pushed the Russian front just fifty miles from the German border, and it was only a matter of time before Russian soldiers crossed into the country. In the west, Allied troops had liberated France and were pushing toward Berlin. The feeling that Germany's collapse was imminent bred paranoia, and the country's leaders were trying to hold on to the last vestiges of control to turn the war around.

At Peenemünde, von Braun's team continued working on an advanced version of the A-4 including one called the A-4b, short for A-4 bastard. It was an enlarged A-4 with two sharply swept-back wings 18.5 feet across welded to its midsection inspired by the Luftwaffe's rocket-powered Messerschmitt Me163B Comet interceptor aircraft von Braun's team often saw streaking through the skies. If they could develop a winged A-4 with room for a pilot, the engineers reasoned, they might be about to show up the rocket planes. Designating their version as the A-4b was a means to fast-track this advanced program by folding it under the guise of the high-priority status granted the A-4 program. This strange-looking missile was launched from Peenemünde on January 24, rising straight up about fifty miles before its fuel stores were depleted. It then arced over to begin the aerodynamically assisted gliding descent

its wings were designed for. It was a successful flight until one
of the wings broke off, sending it crashing to the ground. The
flight was, for a new design, considered a success, but it wasn't
one the rocket team expected to build on. They had bigger and
more pressing concerns.

Not one member of von Braun's team could deny that,
having built the first functional missile launched on civilians,
they wouldn't be popular after the war ended. They each
wanted two things at that point: to escape alive and to continue
working on rockets in the future. Since his release from
Stettin, von Braun had been ready to leave Peenemünde on a
moment's notice—he kept a trunk packed full of vital documents ready at all times. He put his staff on the same alert.

Toward the end of January, the rocket men gathered in a
parlor at the Inselhoff Hotel in Zinnowitz near Peenemünde
and turned to their leader for guidance. Von Braun began the
meeting by reminding his men of their common passion for
rocketry and spaceflight. Then he added that even if Germany
lost the war, they had the upper hand because even enemy
nations wanted their knowledge. This gave the scientists
some freedom to choose where they wanted to go, and so
they weighed their options. By that time the V-2s had been
falling on London for months, making Britain an unlikely
nation willing to support continued rocketry work. The
Russians weren't an option either. From what little the men
knew of Stalin, they felt certain that surrendering to Russia
would be trading one oppressive regime for another. None
were keen on France, which had been under Nazi occupation. That left the United States, a nation that hadn't been
directly affected by the A-4 and that might be interested in
bringing the technology into its own military. They all
agreed that surrendering to the Americans was their best
hope for survival and a future. They just had to find the
Americans before anything happened to them.

Unknown to the German rocket team, the Americans were
on their way to meet them. Weeks after the rocket team's

meeting at the Inselhoff Hotel, the Army Air Force finally gained permission from Eisenhower for the SAG to enter Europe, question German scientists, and inspect German laboratories. The knowledge gained in Europe would feed into the army's Project Hermes, an ordnance program established to first unpack the technology of the V-2 rockets and then build an American version that could meet the needs of future army field forces.

But a meeting between the German rocket team and the American scientists wouldn't be easy to come by in a war zone. The conclusions of the Yalta Conference promised to complicate matters as well. Between February 4 and 11, 1945, President Roosevelt, British prime minister Churchill, and Soviet premier Stalin conferred at Yalta in the Crimean Peninsula of Russia to discuss the fate of Germany. The Allies' intent was to destroy German militarism and Nazism and assist the liberated population. To this end, they decided to break Germany into four occupation zones following the war, each under British, American, Soviet, or French control. Though each nation had its own goals to recover certain people or places after the war's end—the V-2 scientists and sites being high on everyone's list as a form of intellectual reparations—no nation could recover anything or anyone from another's zone once the occupation took effect.

On January 12, 1945, Dryden was still studying the German rockets in the United States when the Russian Red Army mounted an offensive against the German front in Poland. They broke through the Nazi guard and began marching toward Berlin, passing so close to Peenemünde that the rocket team could hear artillery fire. Von Braun knew that as soon as Berlin fell the war would be over, and hearing the war so close by underscored the reality that his time at Peenemünde was fast coming to an end. But for the moment, von Braun and his men remained at the mercy of higher authorities who would issue orders and check they were being followed. The chaos seeping through Germany offered a silver lining. Von Braun knew someone would eventually slip up, issue a conflicting order or something just vague enough to give him a chance to get his men out from under Nazi rule. He would have to be ready.

Unfortunately, von Braun also had to contend with Kammler, the special commissioner for the A-4 program whose recent promotion to lieutenant general put him in charge of the army's V-2 arsenal as well as the scientists who built it. As hope for a German victory faded, Kammler's devotion to the Reich turned into a desperate fanaticism that shone through in the orders he gave von Braun. Kammler first ordered that the rocket team evacuate Peenemünde and hide in Bad Sachsa, a town southwest of Berlin. While his men were readying to leave the site that had been their home of more than a decade, Kammler ordered von Braun to meet with him at the A-4 factory at Mittelwerk. There, in the shadow of the death camp that was still producing deadly missiles, Kammler ordered von Braun to concentrate all secret weapons projects, including the A-4, in the centrally located Harz Mountains near Bad Sachsa. The area was a perfect hiding spot, naturally protected, but also one that kept the rocket team firmly under the thumb of the SS.

Kammler's orders weren't the only ones landing on von Braun's desk. There were others telling him and his men to stay where they were and fight any incoming soldiers to the death to save Peenemünde, their materials, documents, and rockets. It was a wholly terrifying prospect. His men were scientists, not soldiers trained in hand-to-hand combat. Staying behind to fight and protect Peenemünde was a death sentence, but it was also the opportunity von Braun had been waiting for. Conflicting orders gave him some freedom to pick which ones he wanted to follow, and it was lucky that those coming from the SS were those that put his team on the path toward finding the Americans. Disobeying Kammler's orders was the likeliest way to be killed by a firing squad, and evacuating toward central Germany was the Peenemünde team's best chance and the only one von Braun might get. He was going to take it.

CHAPTER FOUR

Escape and Surrender

Allied bombers roared through German skies night and day, each attack on a city, a factory, or a communications center reaffirming that it was only a matter of time before the country was left in ruins and forced to surrender. This backdrop bred chaos, and Peenemünde was no exception. The rocket team could hear Russian guns in the distance. They watched as refugees marched wearily across the island of Usedom near their rocket facility, carrying with them scant possessions and stories of brutality suffered at the hands of a vengeful Russian army. In a last-ditch effort to protect the site, the staff at Peenemünde were armed and the civilian workers trained with the Home Guard in anticipation of a battle for control of the rocket center.

But the chaos helped provide a cover for von Braun's evacuation of his rocket team, as did a typo. Months after the 1943 Allied raid on Peenemünde, Walter Dornberger was transferred from the army's Ordnance department and promoted to army commissioner for special tasks, abbreviated in German to BzbV Heer. The promotion gave Dornberger more power within the army, though it did little to help him counter Heinrich Himmler's attempts to control the A-4 program. A letterhead issued to Dornberger as BzbV Heer had been misprinted as VzbV. Von Braun's transportation coordinator, Erich Nimwegen, who was known to routinely skirt the line of illegal activities, exploited this typo by turning it into the acronym for a fictitious top secret project within the SS, Vorhaben zur Besondern Verwendung, roughly "Project for Special Disposition." With trigger-happy SS guards everywhere, von Braun knew that his team moving documentation and equipment would draw their attention. That no one would recognize the acronym "VzbV" would serve as a shield since the average guard wouldn't have clearance to know about a special project and would know better than to question the team. But to be safe, Nimwegen also wrote transportation orders for the group on the misprinted forms. Von Braun signed them using his rank as a SS Sturmbahnführer, or major. It was one of the few times, if not the only time, he exploited this title.

As the group prepared to leave Peenemünde, "VzbV" began appearing in massive letters on boxes of documents and materials. The acronym was similarly printed in several foot-high letters on the sides of the cars, trucks, and railcars into which the men loaded their material and personnel. The first train, carrying 525 members of the rocket team and their families, as well as boxcars of materials, left on February 17 under the auspices of the VzbV. Almost immediately they were stopped at a roadblock where SS guards were looking for deserting soldiers or civilians shirking the duty of fighting to save their homeland. Nimwegen flashed the travel orders signed by SS major von Braun and pointed to the cars stamped with "VzbV." Rather than risk the punishment of denying an SS special group passage, the uncertain guards stepped aside to let the rocket team pass.

Two days after the evacuation began, the last rocket launched from Peenemünde. A little over a month later, on February 27, von Braun made his final trip to the island center after a month traveling around the country setting up the new site for the rocket team. He told the few men left at Peenemünde that he had found space for them at a central development location. They would even have test stands for the A-4 program, a surprising show of support for their continued work so late in the war.

Von Braun knew it was all in vain. But he also knew the Gestapo was watching his every move, and that he could only retain the logistical support to evacuate his men as long as he maintained the appearance that he was still working on developing the A-4 for Germany's ultimate victory over the Allies. He traveled enthusiastically through the Harz Mountains seeking out mines, schools, and factories that might be used as temporary A-4 facilities. He presented to Kammler plans for new test facilities and launch sites. But von Braun knew no new site could possibly support the program as Peenemünde had, not to mention it was impossible for work to continue uninterrupted with the entire team moving in a slow convoy through the countryside. For von Braun, continuing to organize the program that would build rockets that would never fly was all a charade. Kammler, still convinced Germany could win the war, believed the ruse.

As winter turned to spring, von Braun continued to travel throughout the country under the auspices of advancing the A-4 program. And with a nearly constant overflight of Allied Thunderbolt and Mustang fighter planes, he began traveling increasingly by night to avoid detection, a relative quiet he welcomed in an otherwise hectic schedule. In the early hours of March 12, the rhythmic thumping of tires traveling at sixty miles per hour on the autobahn from Thuringia to Berlin lulled an exhausted von Braun to sleep in the passenger's seat. The sound had a similarly soporific effect on the driver, and near the town of Weissenfels the vehicle careened off a forty-foot-high embankment. Inside the airborne car, the sudden absence of the sound of tires on pavement woke von Braun, who instinctually tried to protect himself from the impending crash by raising his left arm above his head. An instant later, the car landed 130 feet from the road near a railway track. The force was enough to break von Braun's arm and shoulder and knock him and his driver unconscious. Behind them, another car full of men from Peenemünde spotted the wreckage off the road. Von Braun woke up in a private room in a hospital in nearby Bleicherode with his arm in a heavy cast. He was bedridden, his hectic travel schedule replaced by a stream of visitors day and night.

Confined to the hospital and increasingly anxious for the war to end, von Braun learned on his thirty-third birthday that the Allies had crossed the Rhine River. American, Canadian, and British forces had moved from Marseille in France and pushed through Holland so quickly that German defense forces were caught off guard. One armored division participating in the offensive successfully secured the

Ludendorff Bridge before the German defenses could destroy it, preserving a key path toward Berlin.

When the news reached U.S. Army Lieutenant General Omar Bradley, he called Eisenhower, who was at dinner in Reims with division commanders of the Army Air Force. Eisenhower was incredulous when Bradley said he had secured a permanent bridge over the Rhine. It was a dream scenario. Eisenhower told Bradley to send as many divisions over that bridge as he could. Bradley agreed, telling the general that had been his instinct, too, but he'd thought it best to check with the commander of the European Theater first. It was one of Eisenhower's happy moments in the war. The assault had gone like clockwork and the Allies had pierced Germany. By the end of March, German resistance on the Rhine had collapsed. It was only a matter of time before the war would be over. All that remained was for the Allies to take Berlin. Then would come the division of a defeated Germany as per the arrangement laid out after the Yalta Conference.

This politically ordained division of Germany complicated matters for the Army Air Force's Science Advisory Group. Under Theodore von Kármán, the men were planning a trip to Germany to interview the scientists who had developed the A-4 and other advanced weapons and also bring back hardware to the United States as part of Project Hermes, the American V-2 program aimed at developing a new missile. But according to the new political boundaries crossing through Germany, Peenemünde was slated to fall under Soviet control. The Americans would have to reach the rocket site before the postwar occupation went into effect and do so without angering their temporary Soviet allies.

Von Braun, meanwhile, did not know that imaginary lines were being drawn through his country that might prevent his men from reaching American forces. And Hitler's scorched earth policy complicated things for both parties. On March 19, the Führer ordered that all military, industrial, transportation, and communications facilities be destroyed by the retreating Home Guard. He wanted nothing, from documents to hardware to personnel, to fall into Allied hands. Better the Allies inherit a completely destroyed nation than learn the Reich's technological secrets.

News that Allied troops were closing in on Berlin spooked Kammler. He canceled all V-2 field operations effective April 1 and ordered the rocket team to move from their temporary lodging at Bad Sachsa and Bleicherode near the Mittlewerk factory toward Oberammergau, a picturesque village in the Bavarian Alps near the Austrian border. Still bedridden, von Braun began to wonder whether he and his men were pawns in Kammler's game, whether he planned to use them as a bargaining chip, handing the rocket team to captors in exchange for his own life. It was a frightening idea, though not impossible. Most of the team traveled south by train. Von Braun, whose large cast would prevent him from diving off a train and hiding in the event of an airstrike, traveled by car with Dornberger.

Two other men from the team also traveled toward the Alps separately. Dieter Huzel, an electrical engineer and von Braun's special assistant from Peenemünde, was at the wheel of a truck. Seated beside him was Bernhard Tessmann, a test facility designer. Several hours after sundown, the men stopped the truck at an abandoned mine in the Harz Mountains near the village of Doernten. In the back of the truck sat a group of men who could honestly say under interrogation that they did not know where they were. Packed in with the men were fourteen tons of documents. Huzel let the men out, and they all began loading the boxes onto a flatcar parked on a short track that ran into the mine. When the truck was empty, Huzel drove away and recovered a second truck of materials. When that one was emptied, he drove a third truck to the mine.

By eleven o'clock the following morning, the contents of all three trucks were behind a heavy iron door in a room twenty-five feet wide and twenty-five feet deep with twelve-foot-high ceilings one thousand feet inside the mountain.

Huzel and Tessmann alone returned the following day to see that the mine's caretaker had dynamited the tunnel closed, blocking the path to the vault with a jumble of rock and debris. The former Peenemünders left before the caretaker detonated a second stick of dynamite for good measure. They were the only three men in the world who knew that anything lay buried deep in the abandoned mine.

Von Braun and Dornberger arrived at Oberammergau to find the rocketeers living amid stunning scenery in army barracks surrounded by a wall of SS guards and barbed wire. The same day, the U.S. Army's Third Infantry took Nordhausen in central Germany. A week later, the Soviets began their assault on Berlin. The Reich was fast collapsing, and the rocket team was sitting idly in Oberammergau, effectively being held as prisoners. A delirious Kammler arrived on April 11 and announced to von Braun in all seriousness that he intended to singlehandedly win the war for Germany. He unceremoniously transferred command of the rocket engineers to one of his men, Major Kummer, then left brandishing a pistol. It was the last time von Braun, or anyone, saw the man alive.

Kammler's last act of haphazardly transferring control to Major Kummer gave von Braun an opportunity to leverage one man's fear into his men's freedom. He approached their new keeper with Ernst Steinhoff, the former head of Peenemünde's guidance and control lab, and the pair explained the severity of their current situation. Allied air forces were flying over Germany all the time, crushing Germany's industrial war manufacturing capability and its will to fight. With the whole of the Peenemünde team, their work, and their materials in one place, a single well-aimed strike could wipe the A-4 program off the face of the Earth. And if Kummer managed to survive, von Braun told the major, responsibility for the loss of Germany's greatest technology and scientific elite would be squarely on his shoulders.

The scientists paused for a moment and watched the soldier's face as he pictured the scenario, before presenting him with an alternative. Von Braun suggested the major could split the rocketeers up and send them to different villages around Oberammergau so the Allies couldn't kill them all off with one attack. As von Braun paused again to let Kummer think, the sound of an Allied plane roaring overhead filled the air, fortuitous timing that helped Kummer make a decision. He consented to split up the engineers on the condition that each group travel and live under the watchful eyes of SS guards. It wasn't ideal, but the arrangement suited von Braun. His team scattered, waiting to be overrun by American troops.

Von Braun's brother Magnus moved with one group from Peenemünde to Weilheim, a town closer to Munich, while Wernher was transferred to a private hospital specializing in sports medicine about thirty miles away in Sonthofen for further treatment of his broken arm. At some point during his rushed departure from Bad Sachsa and rough drive down to Oberammergau, his arm had slipped in its cast, preventing the fracture from healing. A surgeon in Sonthofen reset the broken bones and encased von Braun's arm and shoulder in a new cast. He also offered the engineer large doses of morphine to numb the pain.

Lying in bed drifting in and out of an opiate haze, von Braun often woke from uneasy sleep to hallucinations of SS officers bursting into his room with guns aimed to kill. The continuous sound of distant explosions and battle did little to soothe his addled nerves. On his third day of morphine unrest, he was shaken awake by a very real armed man in a uniform. His initial panic vanished when he recognized the uniform as a friendly one worn by a messenger sent by Dornberger, who was staying in a nearby hotel. The French Army was just a few hours away, the man said, and the Peenemünde team was leaving. Von Braun's surgeon reluctantly made his patient a hard cast designed to keep his arm bent in a crook while raised at shoulder height so the joint would stay in a straight line. It was cumbersome but allowed the engineer to travel. By nightfall, von Braun had arrived at Haus Ingeburg.

Haus Ingeburg was like an oasis of calm in the frantic final stages of the war, a ski resort nestled high in the mountains above the countryside on the old German-Austrian border. There was little for the rocket team to do but sit on the terrace overlooking the strikingly snowcapped peaks of the Alps against the clear blue spring skies while hotel staff catered to their every desire. The men dined on gourmet meals expertly prepared by the hotel chefs and drank from the stocked wine cellar. They tracked the war's denouement on the radio. They learned that French troops were to the west of them and American troops were to the south, that food supplies were dwindling nationwide, and that liberated concentration camp prisoners and those sent on death marches alike were dying by the thousands, some at the hands of the Nazis and others from starvation.

The final battles were being fought all around the rocket men who remained protected by virtue of their isolated mountain hiding spot. They were safe but stagnant, which frustrated von Braun. He had survived the explosions that came with developing the operational A-4, had endured arrest and imprisonment at the hands of the Gestapo, had survived the raid on Peenemünde, had kept himself alive under the thumb of the fanatic Kammler and walked away from a nearly fatal car crash with only a broken arm. Now, it felt like extreme isolation could be his undoing. All access points into Germany were effectively blocked by Allied soldiers, and American and Soviet troops had met at the bank of the Elbe River near Torgaun, just seventy-five miles south of Berlin. As soon as advancing armies reached Berlin the war would be over. With attention focused on the German capital, it was unlikely American troops would stumble upon his hiding spot in a secluded mountain resort. If American soldiers weren't going to find him, von Braun would have to go find American soldiers.

The rocket team was still ensconced in their mountain retreat at Haus Ingeburg on April 28 when the Army Air Force's Science Advisory Group arrived in Europe. Under Operation LUSTY (for Luftwaffe Secret Technology), the small cohort of seven scientists dressed in military uniforms arrived in London on a C-54 transport plane. The uniforms were little more than costumes to camouflage the scientists and expedite their entrance into Europe. Von Kármán wore a general's uniform while Dryden and the other members bore the rank of colonel. But because the war was still raging in the European Theater, the American scientists were put into a holding pattern, traveling to Paris where they would wait for the war to come to a resolution before embarking on their journey through Germany.

They didn't have to wait long. On April 30, 1945, Hitler took his own life in his bunker in Berlin. The Peenemünde team didn't learn of Hitler's death until May 1, when a propagandist radio broadcast reported that the Führer died valiantly in combat against the Russian Army. Hitler's death freed Germany's soldiers from his grasp, though this wasn't necessarily a positive development for many. Men like Dornberger, whose allegiance to Germany had forced their allegiance to Hitler, were free to fight for their own lives without being labeled as traitors to the Reich. But now they faced the potential of being labeled war criminals for having worked within Hitler's war machine.

The day he learned of Hitler's death, von Braun approached Dornberger and proposed they send someone out to look for American soldiers. Their hiding place, he had learned, was in an area soon to be occupied by French troops, a group they didn't want finding them. Dornberger agreed, and the rocket men staying at the mountain resort convened after breakfast to concoct a plan. They elected Magnus von Braun as their emissary because he was the youngest of the group, spoke the most English, and was the most expendable. The next morning, May 2, the Second World War was declared over in Europe, and Magnus left Haus Ingeburg on a bicycle. He rode down the snow-covered mountain roads toward Austria where reports said American troops had last been seen. He'd gone less than two miles when he found an American anti-tank platoon of the Forty-Fourth Infantry Division. The Americans had no reason to shoot the young German scientist who was dressed in civilian clothes. It took Magnus more than half an hour of German and broken English to convince the Americans that he wasn't insane, that he wasn't trying to sell his own brother for money, there really were 120 of the scientists behind the V-2 rocket camped out in a nearby resort. He asked the soldier to check with his intelligence people, who would surely know rumors of the science team's whereabouts.

From there, everything changed. The Americans recognized the bounty that had fallen into their hands and became quite friendly to Magnus. He was escorted sixteen miles west to Reutte where the Forty-Fourth Infantry had set up a command post. Magnus repeated his story to First Lieutenant Charles Stewart of the Counter Intelligence Corps. The rocket team wanted to surrender right away, he said, before the trigger-happy SS soldiers shot them. And they wanted to start building rockets for America. Stewart listened to Magnus's story and after conferring with his soldiers and intelligence officers, gave him passes that would let him safely cross through the American-occupied countryside back to Haus Ingeburg with orders to return with Wernher and a sampling of the other scientists in tow.

The anxious tension that had permeated Haus Ingeburg throughout the day was broken when Magnus returned slightly after two o'clock in the afternoon. After he breathlessly shared his story, the men decided who among them would go to meet the Americans. Magnus topped the list followed by Wernher and Dornberger, two of the most sought after scientists in Germany. Dornberger's chief of staff, Hebert Axster, and a specialist from the A-4's engine production program, Hans Lindenberg, were also included. Rounding out the group were Wernher's special assistant, Dieter Huzel, and test facilities director, Bernhard Tessmann, the two men who had buried the bounty of documents from Peenemünde in the Harz Mountains. It was a powerful selection of rocket men, a group that could go a long way toward proving to the Americans that they were in fact the famous scientists they claimed to be. The seven men packed into three cars with their luggage and started down the mountain pass around four o'clock that same day. As sleet began falling, darkening the already gray afternoon skies, the men hoped it wasn't an omen. The Americans had received Magnus fairly well, but no one could be sure how they would react with the seven scientists in their midst.

It was dark when the convoy of rocketeers arrived in Reutte. They were greeted by American guards who accepted Magnus's passes and let the rocketeers through. From there the scientists were escorted to Lieutenant Stewart, who greeted them warmly with a meal of eggs, white bread, butter, and the first real coffee they had had in months. Despite the ski resort's gourmet kitchen, the Peenemünders had been drinking ersatz, a sort of coffee substitute, for weeks. The Germans stayed the night in the Americans' site and shared another meal of eggs and real coffee in the morning.

Von Braun was pleased yet unsurprised by the friendly reception by the Americans. He knew that his team and their A-4 technology together was a coveted prize at the end of the war, and if the Americans wanted this technology, which he was sure they did, then it was in their best interests to keep the Germans happy and healthy. Von Braun knew he was far more valuable alive and cooperative than jailed or dead. For their part, the Americans could hardly make sense of the apparent hierarchy among the Germans. Dornberger, the oldest and highest ranking officer in the group, was so reserved he was nearly silent. He struck the American soldiers as a defeated general who had fought on the losing side of two world wars and had a potential lifetime in prison for war crimes to look forward to. Von Braun, meanwhile, Dornberger's thirty-three-year-old subordinate, seemed to be the group's leader. Jovial and heavy after the combination of rich hotel food and forced inactivity due to his broken arm, he just didn't seem like the brilliant leader of the V-2 program that he claimed to be. They couldn't decide whether von Braun was the Reich's top scientist or its most accomplished liar. He was outgoing, excited, and eager to share his knowledge of rockets with the Allies, and happy to pose for pictures. Celebrity, it seemed, suited von Braun, bringing out a natural salesman-like quality in his demeanor. But his outward ebullience masked a private worry that he would be seen only as a war criminal and weapons builder. He impressed upon the Americans the pride he had in his design, and his hope to use it to explore space.

The next day, Hugh Dryden and Theodore von Kármán traveled into Germany. Their first stop was a lab in the north near Braunschweig. The complex, consisting of fifty-six buildings disguised as farmhouses and camouflaged by the surrounding forest, had been largely destroyed by incoming American troops. But what remained deeply impressed the scientists. It was clear that this had once been a burgeoning laboratory that had taken great strides in ballistics, aerodynamics, and jet propulsion. Dryden pored over what documentation he could find and interviewed a number of the remaining scientists about their work, paying special attention to the Germans' technical data on swept-wing aerodynamics and the human physiology of high-speed flight. It was clearly the first stages of study into high altitude manned flight. But it wasn't the V-2 program or scientists the Americans were after, neither was it the prize of Peenemünde. That site was taken by the Soviets on May 5, who found the labs cleaned out.

Less than an hour from Reutte was the Bavarian ski resort at Garmisch-Partenkirchen where the 1936 Olympics were held. The U.S. Army had turned the site into a makeshift administration building and housing for the more than two hundred Peenemünders who had already been rounded up from their nearby hiding places. Von Braun and the cohort from Reutte joined them on May 7 for the inevitable waiting period that precedes an interrogation. Garmisch-Partenkirchen turned out to be a very busy site buzzing with various intelligence officers all hoping to learn the secrets of the V-2 for their respective organizations. Representatives from the Naval Technical Mission in Europe were there, as were men from the Army Air Force's Operation LUSTY and the British-American

Combined Intelligence Objectives Subcommittee, all tasked with evaluating Nazi technology.

And everyone wanted to talk to von Braun, who stuck close to Dornberger. Together they formed an irresistible pair, their combined knowledge more than enough to trade for their safety. Von Braun did all the talking but Dornberger called the shots, a remnant of the hierarchy that had marked their lives working together for the German Army. The pair focused their energies on appealing to the Americans. In his interviews, von Braun often referenced an imminent war between the United States and the Soviet Union, two countries whose opposing ideologies had been put on hold to vanquish the common German enemy. The best thing the United States could do, von Braun said, was arm itself with his rockets.

It wasn't until his interrogation by the Anglo-American Combined Intelligence Objectives Subcommittee that von Braun showed the allies what he really wanted to do with his rockets. In a memorandum he wrote for British investigators called "Survey of Development of Liquid Rockets in Germany and Their Future Prospects," von Braun detailed a fantastic future where rocket travel was the norm. He imagined winged gliders launching on top of missiles, carrying civilians and soldiers alike around the world in just hours, landing like airplanes on traditional landing gear. And this vision wasn't limited to gliders landing on the Earth. His rockets would be powerful enough to land men on the Moon.

Though Britain had its share of rocket and space enthusiasts, the country on the whole was not interested in a spacefaring future; their focus remained on understanding the technical minutia of the V-2 rockets. But von Braun's futuristic fantasy did strike a chord with his American interrogators. They weren't interested in spaceflight, per se, but they were interested in the rockets that could launch people into space. It was clear to American military leaders that the future of combat lay in advanced technology, and they recognized the value in importing von Braun and his dreams of long-range rockets for a military end.

The wheels to import the German rocket team into the United States were in motion before May was out. The U.S. Army wanted to import at least one hundred full V-2s, parts to build more rockets, along with the scientists who had designed the missile to the United States as part of Project Hermes. The U.S. Army Ordnance's request for V-2s reached Colonel Holger Tofty, chief of the Ordnance Technical Intelligence team in Europe. Tofty in turn assigned Majors William Bromley and James Hamill the task of recovering and moving as much material related to the V-2 program as they could out of Germany as quickly as possible. Time was fast becoming a factor. June 1 was the date when the postwar occupation of Germany would take effect, and the V-2 factory sites at Nordhausen and Mittelwerk were in the Soviet zone. Bromley and Hamill had just days to visit these sites and recover anything of value. The two men hurriedly pressed beyond the imaginary demarcation lines toward the abandoned rocket sites. They recovered and packed up as much material as they could find, filling abandoned railway cars with stray pieces of hardware and loose documents. With no complete V-2 anywhere and no obvious parts lying around, they grabbed anything that looked rocket related and added it to their bounty.

The first train of recovered materials left Nordhausen for Antwerp some three hundred miles to the east on May 22. Another fully loaded forty-car train left the site every day for the next nine days, the last one leaving at nine-thirty in the evening on May 31, just hours before the Soviets were scheduled to move into the area. While Bromley and Hamill were busy at Nordhausen, another stroke of good news reached the American Army. Ordnance officer Major Robert Staver had learned of the trove of V-2 documents hidden in the mine in the Harz Mountains. Determined to recover this material as well, Staver attached himself to Tofty's group and started combing the area. He eventually found the dynamited entrance to the mine, and by the end of May managed to dig out and recover the full fourteen tons of documents. Days after the last documents were removed from their hiding spot, the British started setting up roadblocks in the area. The site was now a British occupation zone.

As May turned into June, von Braun was again stuck in a holding pattern. American forces were busy evacuating documents and materials from the V-2 program, but nothing was being done with the scientists. As far as he could tell, they were no closer to moving to America while what remained of their A-4 project was already on its way overseas.

On Sunday, June 17, von Braun was suddenly told to pack his bags; he was leaving Garmisch-Partenkirchen for Nordhausen. As soon as he arrived, he and his guard joined a team from General Electric and Project Hermes. They began a furious search through the region gathering engineers from the V-2 project. In just four days, they collected a train full of rocket engineers and moved them west toward the nearby villages Witzenhausen and Eschwege, which lay just within the American occupation zone. The scientists were safe from the Soviets, but there were still no formal orders to transport them to America. That order would have to come from Washington.

While they waited safely within the American occupation zone, von Braun started preparing. Even though he didn't have a contract from the American government, he remained convinced that the United States would be where he would be free to pursue rocketry with an eye toward spaceflight. He couldn't promise the same for his men. He wanted to bring at least 520 Peenemünde personnel with different areas of expertise with him to America, but that turned out to be an unrealistic number. He was told he could bring no more than 350. In drawing up his preliminary list, von Braun realized that every single member of his team and their families had a German identity to contend with as well. Most were card-carrying Nazis and many, like von Braun, had joined the SS for various reasons. For the moment the Allies hadn't connected the rocket men with the slave labor used to build the V-2s; they saw the concentration camps as the product of the SS and the scientists as technical experts employed by the state. But the longer he stayed in Germany, the more likely it was that the Americans would start to see von Braun and his colleagues as perpetrators of crimes against humanity. The sooner he could get overseas the better, but his wait seemed to be lengthening.

The Soviets were also seeking the Peenemünde team and took advantage of the Germans' lengthy wait. They broadcast messages by radio and even over loudspeakers attempting to court the scientists, luring them to the Soviet Union with promises of a fast return to rocket work and housing for their families. The calls worked to some degree, a handful of Germans choosing to go east, hoping to stay in or at least close to their homeland. The Soviets were also seeking to court Eugen Sänger and Irene Bredt into moving east instead of west. Neither was on the early draft of von Braun's list of engineers to export to America, nor were they highly sought after by the Americans. Stalin, however, was seized by the idea of a precision bombing system that could destroy an American city within an hour or so of launching. He had read their 1944 report on the antipodal bomber and so desperately wanted the couple working for him that he mounted a significant effort to kidnap the two. The effort was in vain. Sänger and Bredt had quickly fled to France after the war's end in Europe.

Finally, at the end of July, the joint chiefs of staff in Washington finally agreed to bring 350 German scientists into the United States under Operation Overcast. Von Braun's excitement at the high number was dashed when he realized this figure covered all branches of the American military, not just the U.S. Army. He, working with Tofty, had a limit of one hundred men. Von Braun protested, and his final list was accepted with 124 names. Notable by his absence was Dornberger.

On September 12, Wernher von Braun climbed into a jeep and set off with a small convoy that included six other rocket men and an escort of army and GE personnel. They drove from Witzenhausen to Frankfurt for a medical exam before traveling to Paris. Not long after arriving at Versailles at five o'clock in the morning, von Braun made his move official, signing his first contract with the U.S. War Department. It was a six-month contract with the option to renew for another six months, paying 31,200 marks a year to his account in Germany. On the surface, Operation Overcast made it look like the Germans would be in the United States temporarily, but von Braun had every intention of making the move permanent.

Late in the afternoon of September 18, von Braun was one of seventeen Germans who boarded a C-54 transport aircraft. They took off from Paris around nine o'clock in the evening, stopped in the Azores to refuel, then landed in New Castle, Delaware, at two o'clock in the morning on September 20. From there von Braun flew on a C-47 to Boston, then took a ferry ride to Fort Strong, a fortified island in Boston Harbor where Operation Overcast had its processing point. As he turned in a stack of completed paperwork, his arrival into the United States was official.

CHAPTER FIVE

Nazi Rockets in New Mexico

Once home to hunters and agricultural villages, the Tularosa Basin in New Mexico in the southwestern United States was all but abandoned in the mid-1300s after a ten-thousand-year occupation. The climate had become increasingly arid over the centuries, turning salt lakes into salt flats before drought made the land inhospitable. Settlers returned in the mid-1800s, cultivating agricultural plots and establishing small ranches. But new laws regarding private ownership of public land combined with another drought in the 1930s brought the frontier era to an end. In 1942, the military moved in. The British earmarked the site as an ideal place to train pilots overseas because of the wide open New Mexico skies but abandoned the site toward the end of 1941, leaving it open for the American military to move in.

Construction of the Alamogordo Army Air Field, six miles west of the town of Alamogordo, began on February 6, 1942, with the first forces moving in two months later. Throughout the Second World War, the Alamogordo Army Air Field served as a training ground for more than twenty different groups who spent six months learning to fly and operate their B-17s, B-24s, and B-29s before heading into combat.

Questions about Alamogordo's usefulness after the war were quelled in 1945 when another military outpost arrived in the Tularosa Basin. A committee of military personnel, men from the Corps of Engineers, and civilian scientists had surveyed topographical and population maps of the Continental United States and found this remote area adjacent to the White Sands National Monument perfect for a missile range. Construction began on June 25, and the facilities were minimal with living quarters, an administrative space, a building for maintenance and housekeeping, and a building to house all technical aspects of the program. To make the site truly livable, six wells were dug to provide freshwater for those who would live and work there. The White Sands Proving Ground was formally established on July 9, and the first troops arrived a little more than a month later. White Sands proved to be a blessing for the nearby Alamogordo Army Air Field. The laboratory would also support the White Sands Missile Range, a site of the new American V-2 program, Project Hermes. The desert firing range was about to get very busy.

The first recovered pieces of V-2 hardware arrived in New Mexico in August of 1945. Tons of materials packed in three hundred train cars backed up the railway for miles. It took personnel twenty days to unload what turned out to be a very mixed bag of hardware, rocket parts, instruments, documents, and warheads. Some pieces were in good shape; the midsection shells of the rockets in particular needed just minor repairs to torn metals. Other parts, however, arrived broken, damaged, vandalized, or otherwise unusable. Almost all the wooden parts used to house elements in control chambers and rocket fins that had been retrieved from their hiding place in the Harz Mountains had to be replaced. There was evidence of poor workmanship and sabotage from the parts built at the factory at the Mittelwerk factory, too. And not one rocket arrived assembled and ready to fly. The Germans had learned from experience that the longer an assembled V-2 sat in storage, the more problems it developed. This wasn't a setback for the American program. Any rocket that arrived assembled would have had to be disassembled for inspection for safety and quality assurance, anyway. It turned out that only two full V-2s could be built from the three hundred train cars full of material. American Army and General Electric personnel would have to learn to put the V-2 together before any would fly, becoming intimately familiar with its parts and systems in the process if they were going to successfully build an American offshoot.

Rockets had never been favored by the American military, though the technology was familiar nationwide after "The

Star-Spangled Banner" was adopted as the United States' national anthem in 1931. Britain's defeat at the 1781 Battle of Yorktown marked the end of the American Revolution and the nation's beginning as an independent country. But continued interference with the young nation's international trade prompted the United States to declare war on Great Britain in 1812. It was two years before British troops arrived in the former colony, burning the capitol building and the White House before moving on to capture the vital seaport of Baltimore. Fort McHenry, guarding the port, became the target of British warships that fired bombs and iron-cased black powder rockets with incendiary, shrapnel, or explosive warheads. Fort McHenry never fell, and the red glare from the rockets illuminated the American flag through a night of attack. "And the rocket's red glare, the bombs bursting in air, gave proof through the night that our flag was still there."

Rockets similar to those used by the British were adopted by the American military in the mid-nineteenth century. But they were cumbersome, required multiple men to pack the black powder, and the full crews that launched them only had simple guide poles as a means to aim the missiles. The advent of better guns, streamlined bullets, and a stronger integration between soldiers, machines, and a tactical approach to war eventually made these short-range rockets all but obsolete as weapons. But seeing the way rockets developed in Germany during the Second World War changed the American Army's thinking about this technology. That these modern versions demanded minimal manpower to launch, could be directed in flight, and could travel significantly longer distances than artillery shells had inspired the Army Ordnance to create its own rocket branch in 1943, which in turn created Project Hermes.

While recovered V-2 hardware was making its way to White Sands for the army's rocket program, the Army Air Force Science Advisory Group worked diligently to prepare comprehensive reports that synthesized everything its seven members learned in Germany for chief of the Army Air Corps General Henry "Hap" Arnold. Each member covered a specific technology while the head of the SAG, Theodore von Kármán, prepared the general reports that addressed the overall implications of these technological advances on future wars.

The group recognized that nations on both sides had begun the Second World War well armed with conventional weapons, and while both had made significant advances as the war progressed, the V-2 was the standout technological achievement. Even if it was expensive, took many man hours to build, and could only carry a fraction of its weight as an explosive payload, it was still a stunning piece of technology. This one missile combined cutting-edge aerodynamics, structural elements, electronic components, servo mechanics, control units, and propulsion elements. And there was no way to counter a V-2 attack. Once it was in flight, it was invulnerable, flying unhindered until it reached its target. And because it flew supersonically, it was silent and impossible to track, qualities that made it a psychologically powerful weapon as well; fear of a bomb arriving without warning was constant. As for the group from Peenemünde that brought the V-2 to life, von Kármán similarly considered them the most capable missile research group in all of Germany. The operational rocket, it turned out, was far from the Germans' only product. While it might have been the most advanced weapon used during the war, the V-2 was the tip of the iceberg. The SAG had found evidence of far more sinister long-range Aggregate rockets that had thankfully never been built during the war.

From the first proof of concept rocket developed at Kummersdorf West, the A-1, to the operational A-4 renamed V-2, the German Aggregate rockets had all been sleek with an internal guidance system and rear fins for stability in flight. But later rockets in the series, beginning with the A-6, showed signs of change. The Science Advisory Group had found wind tunnel models and drawings of advanced, winged

Aggregate rockets designed to fly supersonically, a design goal that had so far eluded Allied aerodynamics experts. The A-6 and the A-8 were intended to test different propellants and a longer fuselage, respectively. The A-7 looked similar to the A-5, the precursor to the A-4, but had wings and was intended to launch from an airplane on high arcing trajectories that would allow it to gather scientific data on supersonic flight. The A-9 and the A-10 were two long-range bomber systems, early incarnations of intercontinental ballistic missiles designed to reach the United States.

The A-9 was a winged version of the V-2. Wide, sweptback wings running from its nose to midsection turned the simple rocket into a glider intended to fly through the atmosphere rather than above it. It would still launch on a ballistic flight path like the V-2, but with wings it would not follow the same path on its fall back to Earth. The aerodynamic stability afforded by the wings would turn it into a glider, reaching a target much farther away. There were even variants of the A-9 with a pressurized cockpit for a pilot, effectively turning the glider into a manned precision bombing system similar to the one Eugen Sänger had envisioned but never built. The A-10 was a multistage weapons system consisting of an A-9 stacked on top of an eighty-five-ton booster that could generate two hundred tons of thrust. The booster would accelerate the A-9 to thirty-six hundred feet per second before its own rocket engine fired to increase the payload's velocity to eighty-six hundred feet per second. Once its fuel was burned, the winged missile would glide through the upper atmosphere to reach a target up to three thousand miles away, which the German Army had noted was roughly the distance between a western European launch site and a major city on America's East Coast.

The SAG had found some drawings and calculations of the A-10 and evidence that a few experimental A-9s had been built. What they didn't find was evidence that one prototype A-9 had flown, the A-4b that von Braun and Dornberger had launched with moderate success over Peenemünde in January 1945.

It was clear to the Science Advisory Group that if Hitler had given the Peenemünde team priority status and more support earlier in the war, things might have turned out very differently. But as it was, Hitler's shortsightedness about these cutting-edge weapons held the Germans back, much to the advantage of the Allies. Now with the V-2 rockets and scientists in the United States, the Army Air Force could develop the technology to its full offensive potential. Von Kármán recommended that the AAF use the Germans' advanced V-2 spin-offs, the winged and multistage versions, as starting points for its own long-range missile systems.

But it wasn't just the rockets and their powerful engines that were worth incorporating into the American military's arsenal. Hugh Dryden was particularly impressed by the Germans' systems, including the autopilot that guided the V-2 to its target. It was more advanced than anything that existed in Allied research laboratories. It would be just as useful for the Army Air Force to bring this advanced autopilot into its defensive antiaircraft weapons as it would to incorporate it into offensive missiles, strengthening America's overall military position. But more than anything, the SAG's reports underscored the importance of putting resources into developing long-range missiles. Given the technological strides taken during the Second World War, Dryden offered his opinion that the next major conflict could easily start with a single missile strike.

Von Kármán presented the SAG's completed work totaling thirty-three volumes to General Arnold on December 15, 1945. Their findings and advice were distributed throughout the Army Air Force, but von Kármán didn't want his group's work to end there. It was clear, von Kármán said, that if this technological trend continued, the next major conflict would see supersonic airplanes and long-range unmanned bombing systems capable of destroying targets thousands of miles away replacing soldiers fighting in trenches. It would be unacceptable for some advanced technology to take America by surprise in some future war.

A peacetime version of the Science Advisory Group, a civilian group of experts advising the Army Air Force, would ensure both the American academic and industry communities would maintain a national technological cutting edge. Nearing retirement, General Arnold left the decision to his successor, General Carl A. Spaatz, who agreed with von Kármán. One week after the Science Advisory Group's final meeting on February 6, 1946, the groundwork was laid for the peacetime Army Air Force's Scientific Advisory Board (SAB). The thirty-man board met for the first time on June 17 under von Kármán as chairman. Like the SAG before it, this board was subdivided into five panels, with Hugh Dryden again leading the group on guided missiles and pilotless aircraft. It was von Kármán's intention that the SAB be an institution where civilian experts could exchange ideas and guide the AAF without being a hindrance, ultimately leveraging technological developments to prevent future wars, not win them.

While the Army Air Force Science Advisory Group was busily finishing its work for General Arnold, the army was beginning to resurrect the V-2s in the New Mexico desert. Wernher von Braun traveled across the country by train escorted by Jim Hamill, one of the majors Colonel Holger Tofty had charged with recovering the rocket documents from the mine in the Harz Mountains, because the government-sponsored former enemy alien couldn't be left alone. When he arrived at Fort Bliss, the closest army base to White Sands, on October 8, von Braun didn't find the America he'd dreamed of. The hot Texas desert site didn't have the accommodations the engineer had grown used to at Peenemünde, and his reception was far from warm. The base's commanding general was a veteran wounded in both world wars and not pleased to be hosting one German scientist, let alone the dozens that would soon arrive. And though von Braun could move freely about the base, he was forbidden to leave the grounds and had to share a room with the major for security reasons. Making matters worse, he had contracted hepatitis at some point during his exodus and was forced to spend eight weeks in the hospital alongside American soldiers. Von Braun's situation improved when the first in a series of trains bearing his fellow Peenemünde engineers arrived at White Sands two months later, on December 8. Another group reached the desert on January 15, 1946, and the last followed on February 20. Once settled, the Germans were assigned to various rocket-related projects, though only thirty-nine were tasked with getting V-2s flying under Project Hermes.

By contrast, no Germans were working on the Army Air Force's rocket program. At the end of October, the Air Technical Service Command invited industry proposals for a concept study and preliminary design for a surface-to-surface missile with a five-thousand-mile range, a significant advance over the V-2's two-hundred-mile range. One company, Convair, presented the Army Air Force with two missiles, one a subsonic winged jet-powered missile and the other a supersonic ballistic rocket-powered missile. The AAF accepted Convair's proposal and quietly awarded the company a contract, marking the beginning of Project MX-774, a test bed on the path toward an American intercontinental ballistic missile. And while it didn't include German personnel, Project MX-774 did use the V-2 as the jumping-off point with some significant changes. The Convair missile's airframe was notably different. In place of a supporting thick body structure like the V-2 used, the MX-774 employed a radical design that used the missile's thin-skinned, lightweight fuel and oxidizer tanks as structural elements. To keep the missile from collapsing under its own weight, the fuel tank structure was pressurized with nitrogen, making it as solid as the thicker-skinned V-2. Another big difference was Convair's use of a detachable warhead. Separating the warhead after its rocket burned out meant that only this small payload would have to withstand the heat of atmospheric reentry as it approached its target. The MX-774's engine was also advanced compared to the V-2's, which was steered by rudders protruding into the rocket exhaust. The new engine, built by Reaction Motors Inc., could swivel to steer the rocket in flight. These changes

coupled with a simple gyrostabilized autopilot system promised to give the Army Air Force a far more sophisticated
missile.

In January 1946, before the first V-2 rockets were assembled
at White Sands, the Army Ordnance Department held a
conference to discuss possible uses of the rocket beyond
learning how it worked. The V-2 was designed to carry a
one-ton warhead to its target, and while the Project Hermes
V-2s wouldn't be armed they would need something equally
heavy in the nose for ballast. It made sense to replace the
warhead with scientific instruments, turning the V-2 into
an upper atmospheric research vehicle as well as an
exploratory weapons system. One result of this conference
was the V-2 Upper Atmospheric Research Panel, which
would use the German rocket to explore atmospheric
phenomena. Its structure, composition, temperature, and
pressure were not well understood, nor was the atmosphere's
impact on cosmic ray and ultraviolet radiation absorption at
high altitudes or the behavior of sound and shock waves.
The V-2 would help answer these questions, but it had to
get off the ground first.

    With the help of the German contingent, the first fully
German-built V-2 engine came to life in a static fire test on
March 3. A month later, the first V-2 actually flew through
American skies, lifting off the launchpad cleanly before
tipping over to fly erratically as controllers watched helplessly, unable to shut it down. One of the steering jet vanes
protruding into the rocket's exhaust was lost right after
liftoff, transferring the steering load to the missile's aerodynamic fins, which couldn't compensate. They ripped from
the tumbling rocket's fuselage, sending the fully fueled
vehicle careening to the ground where it exploded in a fireball that consumed its full load of fuel. The third V-2
launched two weeks later was far more successful. It flew
thirty-one miles before crashing and creating an enormous
crater in the desert.

Project Hermes gained momentum as more V-2s launched from White Sands. Procedures changed and full systems testing brought increased reliability to the complex missile system. Individual components were tested and calibrated before a launch, and once the rocket was assembled and tested as a unit it was ready to fly; from that point on, nothing could be removed or replaced. As they became more familiar with the rockets, General Electric personnel put more energy into calibrating components to increase the missiles' power. The program progressed and spawned additional variants. The Hermes A1 was originally planned as an antiaircraft missile. The A2 was planned as a surface-to-surface model. The Hermes A3 was a tactical missile designed to deliver a thousand-pound warhead to a point one hundred and fifty miles away, give or take just two hundred feet. But these concepts never got beyond the planning stage. The Hermes II version was designed to test a ramjet engine, one that uses the vehicle's forward velocity to "ram" air into the combustion chamber, eliminating the complex and costly rotating compressor and turbine wheels used by conventional jet engines. Another American V-2 offshoot was the Hermes C1, a three-stage missile with clusters of solid fuel rockets to deliver even larger payloads to more distant targets.

On May 29, 1947, the Americans working in the New Mexico desert got a firsthand look at the power of the recovered German rockets. As the Sun set just before seven-thirty on that cool, crisp Thursday evening, a Hermes II rocket lifted off into twilight skies. It was supposed to fly north toward the uninhabited desert, but the onboard guidance gyroscope failed and the rocket turned southward. Range safety technicians saw it was flying off course but still didn't shut its engine down. Observers watched the rocket, trailing flame and vapor, arc to an altitude of forty miles and fly south over El Paso, Texas, then over the international border. Five minutes after leaving the launchpad, the rocket crashed into a rocky knoll just three and a half miles south of downtown Juárez, Mexico. Traveling at twelve miles per minute on impact, the force of the crash shook buildings in the small town and in nearby El Paso. The shock wave broke windows and stopped the clock in the sheriff's office at seven thirty-two. Flames shot into the air, setting the hillside on fire and generating a thick smoke. Startled residents in both towns flooded their local newspaper offices with phone calls asking what had disturbed their quiet evening. Panic bred as rumors flew around Juárez that an oil plant, an underground gasoline storage dump, or a boxcar full of dynamite had exploded. Personnel from White Sands arrived at the crash site around eight o'clock and found the modified V-2 had left a crater fifty feet wide and twenty-four feet deep in the hillside. Mexican soldiers kept civilians and souvenir hunters at bay while American personnel inspected the site. It turned out the rocket had narrowly missed an ammunition dump where Mexican mining companies stored powder and dynamite. The test was a moderate disaster that could have been far worse.

As the American military pursued its various missile programs, a change in national security brought a change to the service branches. The National Security Act of 1947 created the National Security Council, a group that included the president, vice president, secretary of state, and secretary of defense, among others, who met at the White House to discuss both short- and long-term aims in national security. The act created the Central Intelligence Agency, the peacetime civilian agency modeled on the Office of Strategic Services. The act also merged the War Department and Navy Department into a single organization, the Department of Defense. The new DOD, headed by the secretary of defense, directed the creation of a new Department of the Air Force. After a decades-long connection with the U.S. Army, the air force became its own independent military service with its own stations spread across the country. It retained the Science Advisory Board, which now had a much firmer position thanks to the air force's independence.

The same year it became an independent military branch, David Simons, a young physician with a year-old medical degree in hand, joined the air force. He was assigned to the service's Aeromedical Laboratory at Wright-Patterson Air Force Base in Ohio, but the bulk of his work, he learned, would be done in New Mexico with a scientific version of the V-2 called Blossom. Sixty-five inches longer than the standard V-2, it was designed to break apart in the air and parachute a small nose-mounted capsule back to Earth. The air force's Cambridge Laboratory, which had been the Massachusetts Institute of Technology's test site at Hanscom Field during the Second World War, offered the space in the Blossom rockets' noses to the Aeromedical Laboratory for biomedical experiments, an offer the laboratory readily accepted. Simons was named project officer for Animal Studies in high altitude V-2 flights. Working closely with the project's director, James P. Henry, from the Acceleration Unit of the Biophysics Branch, the pair set out to answer questions of whether a man could survive being launched into the upper atmosphere aboard a rocket.

Among the earliest biomedical payloads launched suborbitally on a V-2 were strains of corn seeds that were later planted and cultivated to look for possible genetic effects from cosmic radiation. But seeds were a poor analogue for a human. What Simons really wanted was to launch a monkey, as close a genetic relative to humans that he could use. He figured that with a pressurized cabin and rudimentary life support system he could send a rhesus monkey into the near-space environment. The flight would be a short one. Following a ballistic trajectory, the capsule would only be exposed to the high altitude region no living creature had ever visited for a few minutes, but it would gather useful data.

On January 13, 1948, the Alamogordo Army Air Field was renamed Holloman Air Force Base in honor of guided missile research pioneer Colonel George V. Holloman. By then the newly named air base was deep in preparations for the first monkey flight. The passenger was a nine-pound rhesus monkey named Albert who arrived at Holloman from Wright Field with Simons. Carefully fitted with biomedical sensors so technicians could monitor his heart and breathing rates, Albert was restrained by straps and inserted into an instrumented capsule mated atop a Blossom rocket. The first primate to ride a rocket lifted off on June 11, but the flight was plagued by difficulties. The biomedical instrumentation failed in flight, and a failed propulsion system stunted the rocket's ascent so it reached a peak altitude of thirty-nine miles, barely getting beyond the stratosphere. The rocket didn't explode, but the parachute failed, and Albert's capsule smashed into the desert. But Albert was dead before he hit the ground. The minimal biomedical data the team gathered before launch suggested Albert succumbed to breathing problems in his cramped cabin before the rocket even left the ground. Nevertheless, the experience gained in preparing Albert for his mission was invaluable and directly applied to the preparation of Albert's successor, a second rhesus monkey appropriately named Albert II.

The second primate's accommodations atop a Blossom rocket were slightly less cramped owing to improved instrumentation and a revised parachute system. It was with high hopes for success that Albert II launched on June 14, 1949. The parachute system failed again, ending the flight in a crash landing, but this mission wasn't a complete failure. The sensors on Albert II's body gathered enough data to tell technicians that he survived launch and the rocket's ascent to its peak altitude of eighty-three miles, confirming the life support system worked. Two more rhesus monkeys died on Blossom flights in that year. Albert III was killed almost instantly after launch on September 16 when the Blossom rocket exploded less than eleven seconds after liftoff. On December 8, Albert IV suffered a similar fate to Albert II, surviving liftoff and returning valuable data before being killed when his parachute failed to open.

The engineering behind sending biological payloads into space was improving, and the understanding of how spaceflight might affect living passengers was becoming clearer, but there was still a long way to go in recovering living payloads from space, and Simons was frustrated by the short duration of ballistic Blossom flights. He needed larger rockets to keep payload aloft higher, but such a rocket was not forthcoming. The air force's more powerful Convair-built missile was no longer a candidate for atmospheric research. After test launches at White Sands ended with mixed results, a common occurrence given the nascent state of rocketry in America, the MX-774 program was canceled in 1949. In its place, development began on a U.S. Navy high altitude sounding rocket akin to the V-2 called Viking that had been contracted to the Martin Company.

None of these new rocket programs were taking advantage of the imported German talent. While the makers of the V-2 provided indispensable assistance in the early stage of Project Hermes, by the spring of 1947 they had been entirely phased out of the program and replaced by American General Electric personnel. The development in part suited von Braun who had never wanted to keep working on the V-2 in the United States. But he also wasn't able to work on the bigger and more powerful rockets he was anxious to build. Instead, he sat idly by while Americans launched an increasingly stale technology. His long-term thoughts, meanwhile, remained firmly off the planet. During his lengthy downtime at Fort Bliss, von Braun penned a technically comprehensive work outlining a manned mission to Mars. His vision wasn't for a small-scale scouting trip to investigate the planet before visiting; *Das Marsprojekt* detailed a seventy-man mission akin to the turn-of-the-century scientific expeditions mounted to the North and South Poles.

The slow development of America's rocket technology frustrated more than the V-2 engineer. General Dwight Eisenhower, now the army's chief of staff, felt the fledgling rocket program wasn't getting the support and priority status it desperately needed. The guided missile field and all related fields like electronics, missile guidance, and supersonic aircraft were still poorly understood, he explained in a hearing before the House Military Appropriations Subcommittee. It was in the nation's best interests to focus research energies into these areas that needed more attention. If the United States didn't devote significant resources to developing this technology into a viable missile now, he believed, the nation risked ruin or defeat to an enemy that was pursuing these technologies, potentially the Soviet Union. But the U.S. Congress was focused on reducing military forces and expenditures in the postwar climate, not sinking more money into any kind of new armaments. Eisenhower had the foresight that had eluded Hitler, that developing missiles as weapons was a worthwhile investment, but he could only do so much to raise the program's priority status. Without an immediate threat to the country, congressional support for the program was lacking, and national resources were urgently needed to rebuild the country more than to develop the weapons to counter a theoretical future threat.

It didn't help the army's missile program that its imported German experts were effectively being ignored. The American War Department didn't acknowledge that it was hosting a number of former Nazi engineers until December 1946, largely because it had taken the better part of a year for all of the men to secure their military contracts. Even then, their freedoms remained severely limited. Having entered the country under Operation Overcast and later Project Paperclip, they could work under the umbrella of the U.S. Army, but they were still not formally recognized by the government as being in the country. And for every government official and industry leader who wanted to tap into the German knowledge base there were just as many who refused to work with such a recent enemy. The American public did not know that German scientists were living and working in the country. Being phased out of the Hermes program only increased the Peenemünde group's sense of isolation. The standing order from Washington was for them to just tinker with their old V-2s and wait until they were needed. The subtext was that they might not be needed until a new war developed that would require the same fast-paced development that had spawned the V-2 or the American Manhattan Project.

A near-term future conflict wasn't an implausible scenario. The uneasy postwar peace risked erupting into a fresh conflict, not between the Allies and old Axis powers but between the West and forces behind the political Iron Curtain surrounding the Soviet Union. Premier Joseph Stalin had agreed to an alliance with the United States against Japanese forces in the Second World War's Pacific Theater on the condition that the Soviets gain a sphere of influence in northeast China after the Japanese surrender. Four years after the war's end, the North Atlantic Treaty Organization (NATO) was founded as a transatlantic security agreement designed in part to contain Soviet aggression and expansion through Asia while simultaneously preventing a renewed European militarism. Signed on April 4, 1949, the treaty made it clear that an armed attack against one cosigning nation would be regarded as an attack against all of them, and retaliatory action fell within the parameters of an acceptable response.

Unfortunately, containment in Asia had not gone well after the Second World War ended. After the Japanese surrendered to the Allies on August 15, 1945, the Soviet Union sent troops into Japanese-occupied Northern Korea. When American troops arrived in the southern part of Korea, the Soviets began cutting roads and lines of communication at the thirty-eighth parallel. Two separate governments were emerging. North Korea had strong Soviet and eventually Chinese support, including Soviet military training and arms for its soldiers. The north refused to participate in a United Nations mandated election and remained under the dictatorial rule of Kim Il Sung while South Korea elected Syngman Rhee as president.

As the new decade drew near, the threat of civil war in Korea that could demand American involvement sparked the U.S. Army's need for new arms, and von Braun was called upon to help his new homeland. The army wanted surface-to-surface missiles, one with a 150-mile range and another with a five-hundred-mile range. The facilities at White Sands were just too small to develop and test missiles this large, so to build these rockets the Germans were going to have to relocate to Huntsville, Alabama. During the Second World War, the Army Ordnance and Chemical Warfare Service had opened two arsenals near Huntsville to produce munitions, bringing a brief but lucrative period of employment and growth to the small town. Deactivated at the end of the war, the Redstone Arsenal was subsequently reactivated in November of 1948, and the Army Ordnance Rocket Center was established there on an interim basis the following February before officially opening in June. At the same time, the newly inactive Huntsville Arsenal, a Chemical Corps installation next to the Redstone Arsenal, was appropriated by the Ordnance Research and Development Division Suboffice of Rockets at Fort Bliss so the influx of personnel heading for Huntsville would have adequate workspace. The Ordnance Guided Missile Center was established at the Redstone Arsenal five days later.

But first, bureaucratic demands called for von Braun to formally begin the immigration process. The Peenemünde group's entrance into the United States under Operation Overcast and Project Paperclip earned them military sponsorship that didn't translate into American citizenship. On November 2, 1949, von Braun and an American officer in civilian clothes took a streetcar to Juárez, Mexico, the city his V-2 had nearly bombed two years earlier. There he went to the American consulate, presented his papers, submitted to the requisite chest X-ray, paid his eighteen-dollar processing fee, and returned back to the border with a stamp on his passport. The whole thing has been prearranged with Mexican and American officials, but the formality was a legal necessity. Finally, after living in the United States for nearly five years, von Braun began the mandatory five-year waiting period before he could formally apply to become an American.

Von Braun left Texas for Alabama on April 10, 1950, and upon his arrival became the new center's director. Here, he was finally given the chance to develop a new rocket that had its roots in the V-2 but ultimately advanced the state of the art of rocketry, though he did inherit some older programs. The army's ongoing Hermes C1 program was also moved to the new Guided Missile Center where it became the research pathfinder for the proposed five-hundred-mile range missile as well as test bed to perfect technologies for the smaller 150-mile range missile the army hoped to develop. What the army wanted to avoid was a crash program to develop a missile quickly. Army chiefs felt there were already enough conventional arms available. Von Braun thus had the time to evaluate his new missile system to ensure upgrades and variants would keep this rocket in the army's arsenal as long as possible. In support of this long-term goal was a union between industry, science, and the Ordnance department that pulled together some of the best minds in the nation without having to establish a large science personnel directorate within the service, something that would have been a long, bureaucratic, and expensive process.

With the German contingent gone from White Sands, the American V-2 program wound down and closed within a year. In total, sixty-seven captured and reconditioned V-2s had launched from American soil. After the twenty-seventh missile, the steering and guidance components and electrical cables were American made, increasing the percentage of successful launches. Thirty-two of the Project Hermes launches were classified as failures, but rarely could the root cause be isolated. Nevertheless, the V-2 Hermes program achieved what it set out to do. It was a research program that returned a wealth of data about the missile and gave American Army personnel, scientists, and engineers valuable experience in handling and dealing with large rockets. However, as a research program, there had been no real attempt within the Hermes project to change or improve on the German components, save for changes to basic components that were necessary to get a particular rocket off the ground. But what seemed to be a pending war in Korea thawed the impasse that had kept von Braun and his colleagues inactive at White Sands. Now in Huntsville, the Germans were going back to work, but they wouldn't be the only ones. Rocket technology was also entering the realm of manned flight in the hands of the air force.

CHAPTER SIX

Rockets Meet Airplanes

North of Los Angeles and the San Bernardino Mountains lies the Mojave Desert. In the arid region, the days can be scorchingly hot and the nights bone-chillingly cold. The beautiful, panoramic sunrises and sunsets are at odds with the violent dust storms that can sweep through the area without warning. For centuries, jackrabbits and coyotes were the sole residents among the low brush and Joshua trees, disturbed only by the occasional lone traveler or family passing through the desert toward the gold-rich mountains. Then in 1876, the Southern Pacific Railroad routed a line through the region. The Santa Fe Railroad followed with another line in 1882 and built a water stop named Rod, adjacent to the immense Rodriguez Dry Lake. Rodriguez was anglicized and shortened to Rogers Dry Lake in the early 1900s when the Corum family arrived and called the desert site home. They set up alfalfa and turkey farms; as more settlers came, the family leased land to homesteaders for one dollar an acre. Before long, the Corums dug water wells, set up a general store, and established a post office. But the family's request to formally name the town Corum was denied by the United States Postal Service; there was sure to be confusion with the existing town of Coram, California. So the Corums reversed the spelling of their family name and Muroc, California, was born.

To Army Air Corps Lieutenant Colonel Henry "Hap" Arnold, this expansive land of rattlesnakes dotted with homesteads was a perfect bombing range. The Rogers dry lake bed is a forty-four-square-mile pluvial lake whose parched clay and silt surface is renewed every year in a cycle of rainwater and desert winds that leaves it as smooth and as hard as glass. To Arnold, the lake bed was a self-repairing runway under reliably clear skies, both of which meant reliably good flight conditions. And the area's isolation from major cities also meant protection from prying eyes.

Early one late summer morning in 1933, two men from the Automobile Club of Southern California and two army personnel in civilian clothing accompanied Arnold on a trip out to Muroc. They arrived at six o'clock and woke the town's one resident, who ran both the general store and the filling station. Posing as members of the Auto Club, the visitors inquired about travel routes in the area. Through a string of profanities and abuse about the Auto Club's chosen time of arrival, the man answered Arnold's questions about the desert environment and land ownership. When the group returned to March Field just east of Los Angeles that afternoon they began searching for titles to the land, most of which turned out to be owned by the U.S. Government. In September, at almost no cost to the American taxpayers and before he had secured the legal title to the land, Arnold established the Muroc Bombing and Gunnery Range, a training site for the Army Air Corps' bombers and fighters.

The flight facilities at Muroc had become permanent during the Second World War. In July of 1942, the Muroc Army Air Base was built to host combat flight crews, and soon B-24 bombers and P-38 pursuit planes were tearing through the desert sky and dropping practice bombs on targets on the desert floor. It wasn't long before more planes arrived. Fast-paced wartime developments quickly over-whelmed the Army Air Force facilities at Wright Field in Ohio and moved to the remote desert, a perfect place to put top secret aircraft through qualification and safety testing. Soon, a second site six miles from Muroc was established, also on Rogers Dry Lake. A wooden hangar and basic facilities were built first, and then on October 1 a turbojet, a Bell XP-59A Airacomet, took off from the lake bed. As pilots put the first American fighter jets through their paces in the desert skies, they found that the expansive and reliably flat lake beds surrounding Muroc offered a welcome safe haven to pilots in distress; if they couldn't make it back to Rogers, one of the smaller surrounding dry lake beds was a life-saving

option. The desert airfield was also a perfect spot to test new
and experimental aircraft.

Toward the end of the Second World War, developments
in aviation had run almost parallel to developments in rocketry, with jet-powered aircraft slowly replacing propeller
planes in combat, though they had appeared late in the war.
The first had been the German Messerschmitt Me-262 that
debuted in July 1942. This had been followed two years later
by the American-made Lockheed P-80A Shooting Star in
January 1944. It was clear that, like rockets, these new jet
airplanes were becoming state of the art and would play a
vital role in future wars. And like any new technology, the
jets brought a host of problems to the forefront, specifically
the problem of air compressibility at speeds approaching the
speed of sound. This was exactly the problem Hugh Dryden
had spent the bulk of his career researching with the Bureau
of Standards in the 1920s and 1930s.

Compressibility was a phenomenon known to scientists
well before it became a problem for aviation and was inextricably linked to the speed of sound. In the seventeenth
century, artillery tests had been done with an observer
standing a known distance from a cannon, measuring the
time delay between a flash of light and the sound of the
cannonball escaping the barrel. These tests revealed that
sound travels at about 1,140 feet per second. But this method
was imprecise, and the figure was disputed until 1943 when
twenty-seven American leaders in aerodynamics met at the
NACA headquarters in Washington, D.C. Among the
attendees were Dryden, Theodore von Kármán, and John
Stack, an aerodynamicist from the NACA Langley Memorial
Laboratory. It was Stack who raised the issue of the speed of
sound as something that would soon become a problem for
aircraft manufacturers building faster vehicles. Without
taking into account all the variables, including the heating
properties of air and density at different altitudes, Stack said,
available data wasn't complete enough to determine the true
speed of sound. Dryden offered a workaround, suggesting
they mathematically round up the average measurements for

a starting value. The committee ultimately agreed and settled on 1,117 feet per second as the speed of sound at sea level where the atmosphere is thickest.

Regardless of altitude, a vehicle traveling at or near the speed of sound through the atmosphere will experience shock waves, a discovery that also predates aviation. Nineteenth-century physicist Ernst Mach studied the supersonic flow of gases using a shadowgraph. His photographs showed a bullet traveling supersonically with a clear shock wave in front of it and another trailing behind it. It was the first physical evidence that sound, a mechanical wave that vibrates the air molecules through which it travels, compresses those air molecules that can't get out of the way fast enough. Mach's research led to the measurements that bear his name; a Mach number is the ratio of the speed of an object traveling through a gas to the speed of sound in that gas.

The same shock waves that Mach's photographs revealed create an extremely unstable environment for an aircraft. It is possible for some parts of an airplane to travel at the speed of sound while others do not. In aviation, the challenges of compressibility were first seen with the tips of propeller blades. Though propeller-driven planes flew well below the speed of sound, the combined movement of a propeller's rotation and the airplane moving forward through the air meant that the tips of the blades moved supersonically. The tips met the resistance of the shock waves, rendering the propeller less efficient and effectively slowing the aircraft. It was clear as early as the mid-1920s that the effects of compressibility could quickly become extremely troublesome and noticeably degrade an airplane's overall performance.

The problem became more complicated as planes became streamlined, started flying faster, and their propeller engines were replaced by jet engines. As the sound waves traveling in front of an airplane built up, the air flowing over the wing reached supersonic speeds before the air flowing underneath the wing did. The uneven shock waves buffeted airplanes, inducing structural failure and loss of flight control, which ultimately claimed pilots' lives. But without a fundamental understanding of the physical features of the air flow that was causing this unstable environment, engineers were stumped on how to solve the problem. Most troublesome was the transonic range, the range encompassing speeds just below and just above the speed of sound, roughly between Mach 0.8 and Mach 1.08 where the buffeting effects of the uneven airflow are most pronounced and dangerous for a pilot.

Engineers suspected that once an airplane was flying supersonically it would be stable, but it would have to reach supersonic speeds first, breaking through the compressed shock waves in the process. This became known as the so-called sound barrier. The myth that the sound barrier is a physical wall in the sky is rooted in a 1935 sensationalist headline. While giving an interview about high speed flight experiments, British aerodynamicist W. F. Hilton used a graphic representation of air drag on an airfoil, an aerodynamic object designed to generate lift, to explain the challenge of transonic flight to the reporter. Hilton showed on the graph how the density of the air molecules building up against the wing shot upward to create what looked like a wall. The next morning's headlines coined the term *the sound barrier*, putting the idea of a physical barrier in the minds of people around the world. Engineers of course knew the sonic wall was just an engineering problem, but it was nevertheless a type of barrier that needed to be broken.

The challenge of figuring out how to fly supersonically was made more difficult by inadequate wind tunnel data because contemporary wind tunnels couldn't replicate transonic speeds. Air molecules build up inside a wind tunnel the same way they do in front of a moving airplane, bouncing off the walls of the enclosed space and distorting the air flow. The data engineers gathered was more or less useless, and without wind tunnel data it would be impossible for designers to figure out how to build an airplane that could safely fly supersonically. Rockets were routinely flying faster than sound, but their trajectories essentially replicated that of an artillery shell. Controlled, piloted aviation was at an impasse until an airplane could break the sound barrier.

It was Stack, the aerodynamicist from Langley, who first spearheaded a program to develop a research airplane to take on compressibility by flying supersonically. By the summer of 1943, he and a small cohort from Langley had worked out the basic design details of a supersonic research aircraft. The Stack design called for a small, turbojet-powered airplane capable of taking off under its own power from a runway before reaching a top speed of Mach 1, the speed of sound. What was more, the aircraft was designed to fly supersonically without compromising its ability to fly safely and steadily at subsonic speeds during takeoff and landing. To gather the necessary data, the aircraft would also carry a substantial payload of scientific instruments for measuring the aerodynamic and flight dynamic behavior at near-sonic speeds. And it wasn't meant to be a one-off investigatory flight. Stack envisioned the aircraft first probing the low end of the transonic compressibility regime before flying incrementally faster and attempting to fly supersonically. For Stack, a regimented and measured approach was best.

While the NACA had the knowledge and means to develop this Mach 1 research aircraft, only the military had the money to foot the bill. Fortuitously, the military was also interested in investigating supersonic flight. It wasn't hard to imagine the benefits of having supersonic fighter airplanes as part of the United States' aerial arsenal, which also meant training pilots to fly supersonically. The prospect of experimental aircraft with military applications led to a culture at Langley where fliers were trained as test-pilot engineers specifically for these research programs.

On the last day of November in 1944, Bob Woods stopped by Ezra Kotcher's office at Wright Field in Ohio. Woods, an aeronautical engineer who had cut his teeth in the field working at the NACA's Langley Memorial Laboratory in the late 1920s, had partnered with Lawrence D. Bell in 1935 to form the Bell Aircraft Corporation. Kotcher, at the time, was a senior aeronautic engineer at the Army Air Force's Air Materiel Command. The men chatted informally before the conversation turned to business. Kotcher told Woods that the Army Air Force was interested in building a dedicated research aircraft with the help of the NACA. It would be nonmilitary, not something to be mass produced. Instead, it would be a small run of three aircraft very specifically engineered to fly faster than the speed of sound and return data in the process. This aircraft would doubtlessly inform the next generation of military fighters, but it would itself be purely a research vehicle. If Bell was interested in building it, Kotcher told Woods, the contract for this experimental plane was his. Woods accepted the challenge on the spot.

The Army Air Force's idea of what this high speed research aircraft should look like differed from what Stack had initially imagined. The most notable difference was the proposed power plant. While Stack wanted the research aircraft to be jet powered, Kotcher was sure the only way to break the sound barrier was to swap the jet engine for a rocket engine. In 1943 he had worked as a project officer on the proposed Northrop XP-79 rocket-propelled interceptor. During his tenure with the program he had learned that the U.S. Army knew that the German Luftwaffe was developing a rocket-propelled interceptor as well, the Messerschmitt ME-163. With contemporary turbojet engines unable to push an aircraft faster than sound, using a rocket engine like the Germans was an obvious solution. The Army Air Force ultimately agreed with Kotcher, and the military specifications for the research plane emerged quite different from the NACA's initial design.

The Army Air Force wanted to launch the rocket-powered plane from underneath a high-altitude bomber to conserve its fuel for the short but explosive powered flight, and it didn't want to take a measured approach. The AAF wanted to make its assault on the sound barrier early in the test program. Because the military was footing the bill, naysayers against a piloted research vehicle didn't carry much weight. The aircraft was given the name XS-1 for Experimental Supersonic, which was eventually shortened to X-1. Its rocket engine would be built by Reaction Motors, the same company building the engine for the Army Air Force's Project MX-774 missile.

With the basic design of the X-1 set by the Army Air Force, it fell to Bell Aircraft to figure out the specifics of the aircraft. For inspiration, engineers looked to a .50 caliber bullet, which they knew left the barrel of a gun at supersonic speeds and flew stably to its target. If a bullet could break the sound barrier in level flight, a bullet-shaped aircraft would be able to as well. The NACA stepped in to help fill gaps in Bell Aircraft's research left by missing wind tunnel data. NACA engineers ran drop-body tests wherein winged, bomb-like missiles were released from a B-29 bomber at thirty thousand feet in a rough replication of the X-1's air launch profile. The data was limited but was enough to help inform the design of a supersonic aircraft. Another stopgap test method the NACA brought to the X-1 program was the wing-flow method pioneered by Robert R. Gilruth, chief of the Flight Research Section. He figured that by mounting a model airplane perpendicular to and at the right location above the wing of a P-51D, the air flowing over the wing would be supersonic in a dive. It was far less precise than a wind tunnel, but it nevertheless gathered high speed flow data. The NACA also ran rocket-model tests wherein models were mounted on rockets that were then fired from the launch facility at Wallops Island on the coast of Virginia. The data from all these tests combined with the existing core of compressibility data the NACA had obtained over the previous twenty years became the basis of the scientific and engineering material Bell Aircraft used to design the X-1, breaking from the Army Air Force's blueprint where necessary.

It was exactly this type of creative aerodynamic problem solving and flight data monitoring that the NACA was known for. Created to improve America's offensive air power, the agency was originally led by a committee of twelve volunteers representing government, military, and industry acting in an advisory capacity reporting directly to the president. Initially tasked with coordinating existing aviation programs, the NACA soon grew large enough that it needed its own research centers. The first had been the Langley Memorial Laboratory, established in Virginia in 1917 and formally dedicated three years later with a ceremony that included an aerial display with a twenty-five-plane formation. The center quickly gained a reputation for finding practical solutions to difficult aeronautical problems, a reputation that spread. Management never had to recruit the best minds in the field; the best minds came to Langley on their own and the workforce expanded rapidly.

By 1925, there were more than one hundred employees working at the Langley Laboratory. Part of what made the site so successful was its facilities, which included some of the best wind tunnels in the world, including a pressurized, or variable density, tunnel. Unlike traditional open wind tunnels, this closed tunnel pioneered the use of compressed air to replicate different atmospheric environments. Running parallel to the early successful tests with the wind tunnels were Langley's full-scale flight test programs. These helped set enduring guidelines and promoted a deeper understanding of the instrumentation needed for acquiring accurate data that could then be correlated with wind tunnel data.

The X-1 made its first ten unpowered gliding flights at the Pinecastle Army Air Field just south of Orlando, Florida, in the first months of 1945. These flights were designed to test the aircraft's airworthiness, handling characteristics, and the process by which it would be air launched, all without firing the rocket engine. Bell Aircraft pilot P. V. "Jack" Woolams flew the plane on behalf of its manufacturer while a small NACA team led by Walter C. Williams collected and analyzed the flight data. With these tests complete, the aircraft was unveiled to the public during an open-house exhibition at Wright Field on May 17, 1946. Painted bright orange so it could be seen in the sky and strongly reminiscent of a bullet, the streamlined plane looked every bit the exotic and futuristic aircraft that could break the sound barrier. Just thirty-one feet long, the X-1 was compact with a small, barebones cockpit toward its nose. Straight, stubby wings jutted out from the sides of its fuselage, which housed two propellant tanks, twelve nitrogen spheres, three regulators for pressurization, and retractable landing gear. Though it housed a state-of-the-art rocket engine, the X-1's complexity was kept to a minimum to eliminate as many variables from its already demanding flight plan.

The second production X-1 was the first to be fitted with a rocket engine in September 1946. It was around this time that operations for the program moved to Muroc; both Army Air Force personnel and a small ground crew from the NACA with Williams in the lead moved to the desert site in anticipation of the first rocket-powered flights.

On December 9, twenty-three-year-old test pilot Chalmers "Slick" Goodlin was nestled in the cockpit of the rocket plane as the B-29 mother ship took off from the Rogers dry lake bed. A former Royal Canadian Air Force and U.S. Navy test pilot, Goodlin had left the service and joined Bell Aircraft as a civilian research pilot in December 1943. He was rumored to be making a small fortune flying the experimental aircraft, and Bell was touting him as a hero, the inevitable first man to break through the sound barrier. Goodlin was among a new breed of pilots that emerged alongside experimental aircraft in the wake of the Second World War. It wasn't enough for a pilot to fly a new type of aircraft, not when there were engineers needing data from each flight. These new aircraft demanded test pilots, men who combined the skill of a flying ace with the fearlessness to fly through dangerous and unknown environments without sacrificing the engineering goals of the flight.

That December morning, at twenty-seven thousand feet, the orange plane fell away and quickly lost altitude before the rocket engine kicked in, propelling the aircraft forward as flame exploded from its tail end. Goodlin was forced backward in his seat as he watched the B-29 falling away behind him. Inside the cockpit, the rocket engine's roar was incredibly quiet. He gingerly tested the controls and paid attention to exactly how the X-1 handled under rocket power. Aware of the immense engine firing behind his seat, he maneuvered his controls with the gentle precision of a surgeon, hitting a top speed of 550 miles per hour before running out of fuel. Now without power, he skillfully brought the X-1 to a gliding landing on the lake bed at Muroc. He came down from that flight knowing what the airplane could do. He was sure it could fly a thousand miles an hour, and was sure that he would be the one to make that flight.

But Bell's public celebration of Goodlin proved premature when the pilot began disputing his fee. Goodlin's original contract with Bell stipulated that he take the X-1 up to Mach 0.8 as phase one of the program. He'd completed this goal. But for phase two, the phase that would see Goodlin push beyond Mach 1, he wanted additional financial compensation in light of the risk he was taking. In renegotiating his contract Goodlin asked for a bonus of $150,000 paid over five years to avoid high taxes. Bell Aircraft's lawyers rejected Goodlin's proposal, and the pilot refused to fly until the issue was resolved.

Fed up with the delays this stalemate over Goodlin's contract was causing, the Army Air Force decided to take over the X-1 program earlier than expected in June of 1947, the same month Goodlin's contract with Bell was canceled. By then, the two airplanes together had made twenty-three powered flights and fourteen glide flights, demonstrating its structural integrity to a top speed of Mach 0.82. The rocket engine had been finicky, but overall proved itself as safe and reliable. The hitch with the X-1's takeover by the Army Air Force was that the plane was suddenly without a pilot.

One afternoon in May before the aircraft's transfer was finalized, all the fighter pilots at Wright Field were called in to a meeting and asked who among them would be willing to fly the experimental rocket plane. Among the eight or so volunteers was Captain Chuck Yeager, who hadn't heard various flight test engineers refer to the X-1 as a death trap owing to the unknowns of supersonic flight. A few days later, Colonel Albert Boyd called Yeager into his office. Boyd, chief of the Flight Test Division, asked the young pilot why he'd volunteered to fly the experimental aircraft. Yeager responded honestly that it sounded like an interesting program, something interesting to fly. Boyd impressed upon the young pilot that the X-1 wasn't just an airplane, it was the airplane that would change aviation. There were airplanes on the drawing board that would fly five times the speed of sound, airplanes that would take pilots into space, and they all hinged on the X-1 breaking the sound barrier.

Though he was one of the most junior fliers at Wright Field, Boyd considered Yeager to be the most naturally instinctive pilot he had ever seen, a pilot who could fly with extreme precision without sacrificing rock-solid stability in the air. Flying the X-1 supersonically would demand these traits, and not only did Yeager have them all in spades, he also had the right personality to remain extraordinarily calm and focused when faced with a stressful situation. In June, Yeager went to Bell Aircraft's facilities to check out the X-1 in person. He crawled around the cockpit and was invited to fire the rocket engine while it was safely held to the ground. The sound accompanying the burst of flame that shot twenty feet from the tail end of the aircraft was so shocking and loud Yeager covered his ears with his hands. It was incredible. He wanted to fly the rocket plane, and Colonel Boyd told Yeager it was his.

Before Yeager's selection as the X-1's new pilot was made public, news that Goodlin would not fly the X-1 supersonically had trickled through the aviation industry and into the academic sphere. It reached Scott Crossfield, a young pilot turned engineer who was pursuing his bachelor's degree in aeronautical engineering at the University of Washington. He wrote a letter to Bell Aircraft to volunteer to fly the research airplane through the sound barrier himself, highlighting his relevant career experience. He was a licensed private pilot and graduate of a civilian aviation school who had withdrawn from university to work for the Boeing Aircraft Company before joining the Army Air Force. He had then returned to Boeing briefly before joining the navy as an aviator. After finishing his flight training, he had served as a fighter and gunnery instructor and maintenance officer before spending six months overseas during the Second World War during which he never saw combat. Home from the war, he had returned to university under the G.I. Bill and joined a naval air reserve unit at Sand Point Naval Air Station, maintaining his proficiency by flying fighter aircraft on weekends. Though Crossfield was just twenty-six, his résumé spoke to his readiness to take on a new challenge like flying a rocket-powered aircraft.

If Crossfield's letter reached Bell Aircraft, it fell on deaf ears and likely landed in a wastepaper basket. Not long after he mailed his appeal, Crossfield read in the newspaper that Yeager would be replacing Goodlin. That he wouldn't have a chance to fly the X-1 didn't dissuade Crossfield. Instead, it strengthened his resolve. He threw his energies into his studies, progressing directly into a master's degree program after finishing his bachelor's, all the while maintaining his unwavering passion for flying. He was sure that his time in the aviation world was yet to come.

The X-1 was just the beginning of a shift in the landscape of aviation. The NACA was changing, too. In September, while Yeager was preparing for his assault on the sound barrier, Hugh Dryden resigned from the Bureau of Standards to take on the role of Director of Aeronautical Research at the NACA. Under the direction of a man who had spent a career investigating phenomena on the cutting edge of aviation, the new leadership promised changes. It was that month that the air force became a separate service branch as well.

An hour before sunrise on Tuesday, October 14, 1947, Muroc Air Force Base was buzzing with activity. Technicians were busily readying the X-1 for a powered flight, fueling the orange aircraft with ethyl alcohol cut with water and liquid oxygen before carefully installing it in the bomb bay of a B-29 bomber. When the Sun finally rose four minutes before seven, the day was revealed as bright and clear with only the occasional scattered clouds. After hours of preparation, the bomber finally took off at ten o'clock into welcoming skies.

Painted on the X-1's side that morning were the words *Glamorous Glennis* in honor of Yeager's wife, who had not been happy when she dropped him off at Muroc that morning.

Yeager had been thrown from a horse a couple of days earlier and refused to see a doctor about the two ribs he'd broken; he wasn't going to let anything keep him from flying the X-1. By this time, he had flown eight powered flights in the rocket aircraft and knew its quirks. It was designed to withstand three times the stress Yeager could, and the only way to prove it could fly supersonically was to fly it supersonically. Yeager was getting tired of these incremental flights approaching Mach 1. If the flight that morning went off without a hitch, he privately promised, he would push the airplane through the sound barrier on the very next run whether the air force thought he was ready or not.

Once the bomber was airborne, Yeager gingerly climbed down the ladder from the mother ship, settled himself in the X-1's cockpit, and closed the hatch with the help of a sawed-off broomstick handle Captain Jack Ridley, the B-29's pilot that morning, had left for him. Only Ridley knew about Yeager's broken ribs. He'd left the broomstick to give Yeager enough leverage to close the hatch without hurting his injured side. Just before ten-thirty, twenty-thousand feet in the air, the X-1 fell away from the B-29. Yeager dropped five hundred feet before getting the aircraft into a nose-down orientation, but once he did he lit the four barrels of his rocket engine in rapid sequence. A trailing exhaust jet of shock diamonds appeared behind the aircraft as it accelerated to Mach 0.88 and began to climb. Now in the little under-stood transonic realm, Yeager shut down two of his rocket barrels and tested the aircraft's controls as the Mach meter showed the X-1 was still accelerating. Invisible shock waves buffeted over the wings as the aircraft reached forty thou-sand feet and Yeager relit one of his rocket barrels. Still monitoring his instruments, Yeager watched in shock as the needle on the Mach meter moved smoothly from 0.98 to 1.02 before jumping to 1.06. Yeager was flying supersonically and he hadn't felt a thing. He called to Ridley in the B-29 that his Mach meter must be screwy because he'd apparently gone right through the sound barrier without his ears or anything else falling off.

It had been exactly as the engineers predicted. The sound barrier wasn't a wall to break through, it was an engineering challenge, and the expertly designed X-1 was stable after passing through the transonic range without serious problems. For Yeager, finally going supersonic after all the anticipation and worry was a bit of a letdown, but the implications for aviation were significant. The flight marked the beginning of a new phase in aviation and the beginning of a split in Muroc's personality. Evaluation flights of new aircraft would always have a place at the desert outpost, but now experimental aircraft would be flying alongside them.

CHAPTER SEVEN

A New War, a New Missile, and a New Leader

In December 1949, Muroc Air Force Base was named Edwards Air Force Base in honor of Captain Glen Edwards who was killed flying a prototype Northrop YB-49 Flying Wing. Predominantly owned and run by the air force, the NACA did retain a presence. In November 1949, the organization had set up a site there called the Muroc Flight Test Unit. The nearly one hundred employees fell under the direction of Walt Williams, chief of the station. The NACA, too, had a new leader in Hugh Dryden, the organization's first ever director. As its senior full-time officer, Dryden managed from headquarters in Washington the activities of the Langley, Lewis, and Ames laboratories, as well as the new Muroc Flight Test Unit that operated under Langley's direction. And as he had done each time he had been promoted within the Bureau of Standards, Dryden used his position to shift the NACA's research efforts toward his own interest in supersonic and hypersonic flight.

As the NACA began a shift toward increased supersonic flight research, Scott Crossfield was completing his master's degree and researching various career paths. Being a research test pilot was his dream career, and the NACA's Ames Laboratory in California was his dream employer. Established in 1939 as the NACA's second field site, Ames was known for its sophisticated wind tunnels, research aircraft, and the academic approach to theoretical aerodynamics its engineers and scientists applied to the leading aeronautical problems of the day. Ames, more than any other NACA site, was ensconced in the academic tradition Crossfield was comfortable with, and its activities appealed to his own academic interests in the field. He submitted his application but the reply came back with an order for him to report to the Muroc Test Flight Unit

at Edwards Air Force Base instead. Crossfield was disappointed. This smallest NACA center had just two small
research pilot groups with two or three pilots apiece and a
handful of engineers, and he knew the base was primarily an
air force site, not a research site.

When he arrived in California for his NACA interview,
Crossfield's worst fears were realized. The Muroc Test Flight
Unit consisted of a single hangar in the sand on the edge of a
runway, and it barely had legs to stand on. Everything, from
its running water to pilots, came from the Air Force. But the
atmosphere inside that hangar was at odds with its shabby
exterior. There was a pioneering spirit among the small group
of men working there. They seemed ready to take on new
challenges, ready to push aviation into new realms and build
the tools they would need along the way. And the spirit was
due in large part to Walt Williams, the former Langley engineer who had moved to Edwards with the X-1 to supervise
the NACA's role in the program.

A man of action, Williams was eager to show Crossfield
the other research aircraft at the site that were still in the early
phases of development. There was the X-1, now modified to
fly faster with the goal of gathering more data sets. Following
in its footsteps, the next generation of research aircraft was
ready to address unanswered aerodynamic questions: rocket
planes that would launch from underneath bombers like the
X-1 but featured new design elements like more powerful
engines and more aerodynamic, swept-back wings. It quickly
became clear to Crossfield that this little hangar at Edwards
was where the future of aviation would be shaped, and he
wanted to be a part of it. When Williams offered Crossfield a
spot at the Murco Test Flight Unit, the pilot promptly
accepted. He returned to Seattle long enough to collect his
master's diploma and resign from his naval reserve unit before
beginning the long drive south in June 1950.

Three weeks later, Crossfield's former naval reserve unit
was deployed to Korea. On June 25, seventy-five thousand
soldiers from the Soviet-backed North Korean People's Army
crossed into the southern Republic of Korea in a coordinated

attack at strategic points along the thirty-eighth parallel on a path toward the South Korean capital of Seoul. To the United Nations Security Council, which had lost its Soviet representative six months earlier over the refusal to delegate a seat to China, the move was an international crisis. American president Harry S. Truman committed American forces to a combined United Nations military effort to back South Korea, a move that technically qualified as police action rather than a declaration of war but still marked a change in America's foreign policy. The president established his Truman Doctrine and the Marshall Plan to funnel aid into Europe and contain Soviet expansion while directing the National Security Council to analyze Soviet and American military capabilities. The subsequent recommendation to the president called for an increase in military funding to suppress the Soviets. This manifested as an increase in funding for missile development programs in the United States as well as a consolidated command structure in Europe.

Where Europe was concerned, Truman had asked Army General Dwight D. Eisenhower to command a multinational military force under the North Atlantic Treaty Organization to protect European nations from a possible Communist invasion. The general accepted the appointment. He firmly believed that NATO was America's best defense against Soviet imperialism and took his post as the supreme Allied commander at the European headquarters in the Parisian suburb of Rocquencourt near Versailles.

More than a year after the conflict began and in spite of Truman's efforts, the war in Korea showed no signs of nearing a resolution. By the end of 1951, American soldiers were dying in droves and Truman's approval rating dropped as the mood in the United States became dismal. The president's term was coming to an end, and the current climate made it unlikely that he would seek reelection. Looking ahead, both the Republicans and Democrats shared the goal of healing the nation though neither was sure how to secure victory in the upcoming election. What each needed was a candidate with the right combination of international political understanding and appeal to the common voter, and both parties singled out Eisenhower as their man. The victor of the 1941 Louisiana Maneuvers and celebrated general who orchestrated the successful invasion at Normandy to begin the Second World War's end in Europe was also a proud nationalist with a winning smile. Just coming up on the end of his first year serving as NATO's supreme commander in Europe, it was clear Eisenhower had the nation's best interests at heart and a superior understanding of how to manage a nation at war. The only questions that remained were whether Eisenhower would publicly declare himself for any one party and whether he would actually run.

Eisenhower had always rebuffed the call to politics, citing army regulations that forbade partisan political activity by serving officers. As an active military servant, Eisenhower could not be seen to be advising any one group over another, and publicly declaring his allegiance to one political party would violate this regulation. But the pressure to serve his country in a political rather than military capacity was fast becoming something he couldn't ignore. He learned that Senator Robert A. Taft was the leading candidate for the Republican nomination, a candidate who didn't support Eisenhower's esteemed NATO. A Republican himself, though still unwilling to declare himself publicly as such, Eisenhower offered Taft a deal: he wouldn't run for the Republican candidacy on the condition that Taft agree to support collective security in Europe. Taft refused, and Eisenhower began to see how the sense of duty he'd felt for his country might have an outlet in politics. President Truman, similarly uncomfortable with the prospect of a Taft presidency, was willing to seek another term in office just to attempt to keep the rival Republican candidate out of office.

His aversion beginning to weaken, Eisenhower authorized his close friend Clifford Roberts to organize an advisory group of trusted men to quietly keep him informed of the ongoing political situation. Pressure to declare himself a candidate was unrelenting, and after months of discussion Eisenhower conceded to allow a draft movement. He would not seek the presidential nomination, he concluded, but would entertain what the public had to say on the matter. Henry Cabot Lodge, a Republican senator from Massachusetts, registered Eisenhower as a Republican candidate in the New Hampshire primary election on January 6, 1952, completely without Eisenhower's knowledge. Demands from the press saw the general give a noncommittal statement. If he was offered the Republican nomination for the presidency, he said, he would accept, but he privately remained unconvinced that he should run. Eisenhower held fast to a stance of nonparticipation and refrained from speaking out on national issues or even acknowledging his candidacy. His own humility bred doubt that interest in his presidency was genuine. He was flattered by his friends' conviction, but ultimately unconvinced that the public would support him.

While his supporters were taking steps to secure his candidacy in the United States, Eisenhower returned to France along with his wife, Mamie, to fulfill his commitment to NATO. On February 11, famed aviatrix and businesswoman Jacqueline Cochran arrived at the Eisenhowers' residence in Paris with a film reel in hand. The movie, *Serenade to Ike*, had been shot three days earlier when leaders of the Draft Eisenhower movement had staged a rally in Madison Square Garden. The venue was packed well beyond capacity with twenty-five thousand crammed into a space designed to hold sixteen thousand, and neither police nor the fire marshal could get a single person to leave. Throughout the crowd were signs proclaiming I LIKE IKE. Watching the film, Eisenhower realized he hadn't been so emotional in years, and when Cochran raised a glass to toast the general as the future president of the United States, he burst into tears. He decided his country needed him. Eisenhower announced his candidacy the next day. One month later he won the New Hampshire primary with 50 percent of the votes compared to Taft's 38.5 percent. It seemed the general's distinguished career appealed to both Democrats and Republicans alike.

On June 2, Eisenhower retired from the military, ending a nearly five-decade-long career. Two days later he began his campaign for president in earnest.

The Republican National Convention held in Chicago the following month saw Eisenhower secure the Republican nomination for president. His Democratic counterpart was Adlai Stevenson, a former lawyer and governor of Illinois who had helped organize the United Nations before serving as an adviser to its first American delegation. His candidacy secured, Eisenhower's team chose Senator Richard Nixon from California as his running mate. Nixon was a young man so a counterbalance to the presidential candidate's sixty-one years and was a recognizable name whose relatively short career was backed by strong credentials. Both men were vehemently anti-Communist, which was vital to their campaign. Paranoia that foreign Communist agents were trying to infiltrate the American government was widespread throughout the nation, and the ongoing Korean War weighed heavily on voters' minds as a potential precursor to a feared Soviet invasion of the United States. The U.S. Army at that point relied heavily on the draft to fill out its ranks, which brought home the reality of war. The nation as a whole wanted the conflict in Korea settled, and this was foremost among Eisenhower's promises. He pledged to personally travel to Korea to meet with leaders there and seek a resolution. His only means to serve his country was to bring peace, he said, and the promise exhilarated the nation.

On November 4, 1952, Americans went to the polls. Eisenhower spent the day painting at his residence at 60 Morningside Drive in New York City. Consensus as Election Day began was that the presidential race would be a close one, but it wasn't. Eisenhower won by a landslide, capturing thirty-nine of the forty-eight states including the typically Democratic Florida, Texas, and Virginia. He secured 442 electoral votes to Stevenson's 89. It was a personal triumph for Eisenhower that also pleased the nation. When he was sworn in as president on January 20, 1953, his approval rating was close to 68 percent.

When Eisenhower was elected as president, the state of missile technology was taking strides forward, though the pace had been conservative since the outbreak of the Korean War. Upon his arrival in Huntsville, Wernher von Braun had been tasked by the U.S. Army to develop a missile, or at least a functioning prototype, within thirty-six months. The Hermes C1 missile that had been transferred to Alabama with the German scientists became the basis of the army's long-range ballistic missile. Originally envisioned as a three-stage rocket capable of delivering a thousand-pound warhead to its target, it came to life slowly at the Redstone Arsenal with the idea that it would be built entirely in house with twelve missiles ready for testing by May 1953. Under von Braun's leadership, this program was granted access to the best facilities in the country including government wind tunnels for aerodynamic research and proving ground facilities, as well as adequate financial support and personnel numbers. Technological developments included a new, more powerful engine that, like the missile itself, was based on the V-2.

But the in-house nature of the program at Redstone didn't last. Army Ordnance determined that the research and development site would remain a research and development site and not spend its time building rockets. The legwork of producing components could be farmed out to external contractors and industry partners while assembly and testing of the Redstone missile—the Hermes C1 was renamed after the arsenal on April 8, 1952—would be done in Huntsville. Where the contractors were concerned, the army opted to forego aircraft contractors that would inevitably favor air force programs and look instead toward the automobile and locomotive industries. The Chrysler Corporation was among the candidates, and when its project to develop a jet engine for the U.S. Navy was canceled the company suddenly had the available personnel and facilities to bring the Redstone to life. Chrysler won the prime contract for the new rocket on September 15. Among the subcontractors were North

American Aviation, who would build the engine; the Ford Instrument Company, who would design the guidance system; and Reynolds Metals, who would build the rocket's fuselage. The idea behind contracting out the pieces of the rocket had the Germans put to better use than they had been at White Sands, developing new systems rather than assembling and launching existing rockets.

The development of weapons was keeping pace with the missiles that would eventually launch them. On November 1, 1952, the United States successfully detonated its first hydrogen bomb on Eniwetok Atoll in the Pacific Ocean. But on the whole both programs were progressing slowly. In early 1953, Eisenhower was in the White House, and the new president's administration turned to the nation's missile experts to determine how quickly a working system could be brought to fruition. The successful hydrogen bomb detonation had given the United States a short-lived advantage in the nuclear arms race over the Soviet Union who, on August 12, 1953, successfully tested its own hydrogen bomb. Though smaller than the American one, the Soviet's was lighter, meaning it was better suited to being launched as a warhead on a missile. The advent of the hydrogen bomb pushed the American missile program forward, though the threat these weapons were addressing was unknown. Because the Soviet state was a closed one, there was little solid knowledge about that nation's missiles, military capability, or intentions. Without clear knowledge of Soviet targets, American military strategists could only plan in the abstract and attempt to protect the country from a surprise attack. At every turn, any misstep could spark a new international conflict.

What the United States wasn't lacking was missiles to choose from. With multiple programs running simultaneously, a committee headed by the president of the Massachusetts Institute of Technology, James Killian, surveyed the options. The army's Redstone missile program was one, though the service had others to offer. Jupiter was a longer-range offshoot of the Redstone. In the early 1950s, fifty-seven Redstones were designated as test missiles, seven of which were never flown. Thirty-seven missiles were launched as part of the research and development program, testing new technological developments from new engines, airframes, and guidance systems. Of the thirty-seven missiles, only twelve were part of the Redstone rocket's development. The rest were used to test components that would eventually make their way into the Jupiter missile. These first generation missiles designated Jupiter A were designed to gather design data, test the guidance system, and evolve separation procedures for multistage missiles among other technical goals. Three modified Redstones were designated Jupiter C for composite reentry vehicles that would test a scale module Jupiter nose cone along a specified trajectory to duplicate the reentry conditions the full-scale model would face. The air force presented other options, one of which sprung from the MX-774 program. This early missile effort had been canceled in an attempt to cut military spending in 1947, but the long range missile idea was rehashed in 1951 as Project MX-1593, nicknamed Project Atlas.

Killian's committee recommended that the United States develop more than one rocket with highest priority granted to the air force's Atlas and Titan missile program. There were two corresponding intermediate range missile programs that were also funded, the air force's Thor missile and the army's Jupiter missile.

In 1954, a new group joined the air force missile effort. The Western Development Division was set up in Los Angeles, California, with the task of building a missile that could travel five thousand miles, roughly the distance to the Soviet Union from the east coast of the United States. For some, this rocket brought with it a secondary application as a space-launch vehicle, but this vision was quashed by Secretary of Defense Charles Wilson and his secretary, Don Quarles. These civilian overseers of the military project didn't share the lofty visions of spaceflight and appropriated too little funding to see these projects come to life in the immediate future.

The influx of funding and attention paid to missiles inevitably gave way to satellite proposals, though they were firmly secondary interests. Project Feedback surfaced at the end of 1953 as a complement to the air force's Atlas missile program, a small, Earth-orbiting reconnaissance satellite that would double as a scientific platform. A booster eighty feet long, nine feet in diameter, and weighing 180,000 pounds could deliver a satellite about thirty feet long weighing forty-five hundred pounds into orbit. It could circle the planet fifteen times each day, scanning the land beneath it. A second reconnaissance satellite program was proposed to go beyond this first step. Project 1115, the Advanced Reconnaissance System, was approved by the air force in July 1954 with the goal of establishing critical reconnaissance satellite components and determining how to integrate this new technology into the missile-turned-launch vehicle system.

The proposed satellite systems brought a host of technical unknowns to the fore. How a satellite would be powered once in orbit around the planet, for example, was a problem the Atomic Energy Commission was keen to solve, carving out its niche in the new weapons landscape. How to stabilize the satellite for scanning and imaging the Earth was another question. Equally important was the perfection of an advanced information processing system to support acquisition and distribution of data gathered by these proposed satellites. The state of the art was clearly supporting the loftier goals of these new technologies. Into the mix of unresolved questions were the rocket-powered manned research flights flying over Edwards Air Force Base. As they increased in sophistication, it seemed that there might be a place for a man in space as part of some future reconnaissance or military program. Which raised the equally troublesome question of what might happen to a man flying a strange new vehicle outside the safety of the Earth's atmosphere.

CHAPTER EIGHT

Higher and Faster

In 1841, an act of Parliament transferred some twenty-eight acres of land from the grounds of London's Kensington Palace to the Commissioners of Woods and Forests. The idea was to develop the land and turn it into housing for high-profile tenants, the revenue from the lease of which would go to maintaining and improving other royal gardens. The commissioners' plan for the land centered around a broad avenue seventy feet wide called the Queen's Road that would connect Kensington High Street with Uxbridge Road. The remaining land would be divided into thirty-three plots for detached and semidetached houses. Each plot would be leased on a ninety-nine-year term beginning on Lady Day 1842 to those willing to develop houses valued at at least three thousand pounds to attract wealthy occupants. The revenue from this proposal would be significant; if all thirty-three plots were let for the minimum rent, each estate would bring in twenty-three hundred pounds per year. Hopeful lessees eager to develop the land were asked to submit building plans to the commissioners first to ensure they met certain conditions, including a pledge that the house be ready for habitation within two years of the lease being granted and that the house include ornamental gardens, boundary walls, and iron gates that could allow passage for carriages.

Once the British Treasury authorized the commissioners' plan in January 1842, existing buildings on the land were razed and construction began on the Queen's Road. In July, Samuel West Strickland leased the first land, three adjoining plots along Uxbridge Road. The following September, John Marriott Blashfield secured a lease for twenty plots. Days later, he submitted the plans, elevations, and specifications of his first house, number 8, to the commissioners. The architect, Owen Jones, had incorporated a considerable amount of internal and external ornamentation into his design, and when number 8 was completed in 1846 its moresque, garish enrichments and expansive plain front gave it an exotic quality better suited to a Black Sea resort. It was on the whole somewhat at odds with the Victorian and Edwardian mansions that were slowly taking shape along the avenue.

Number 8 remained unoccupied until March 1852 when Mrs. Caroline Murray bought the house for sixty-three hundred pounds. Finding it far too large, she built an extension on the house's south side then divided it into two units and leased the northern half to barrister and Recorder of London Mr. Russell Gurney in 1854. Mrs. Murray and Mr. Gurney were in good company. By the end of the nineteenth century, the area was second to none in terms of attractiveness of surroundings, and, renamed Kensington Palace Gardens, was home to bankers and leaders in the world of finance.

Number 8 Kensington Palace Gardens was in a state of disrepair in July 1940 when it became the headquarters of the London office of the Combined Services Detailed Interrogation Center. Known colloquially as the London Cage, the once opulent estate served as a temporary home for former Nazis and German prisoners of war subject to harsh interrogations at the hands of British officers. The leader of the London Cage was Alexander Scotland, head of the Prisoner of War Interrogation Section of the Intelligence Corps. It was here, into Scotland's hands, that Walter Dornberger was sent like a lamb to a sacrificial altar just weeks before Wernher von Braun arrived in the United States.

Von Braun and Dornberger both had arrived at the Bavarian ski resort at Garmisch-Partenkirchen in May 1945 for interrogation by the Allies. But while von Braun was taken by the Americans and asked to draft a list of men to accompany him to the United States under Operation Overcast and Project Paperclip, Dornberger was among eighty-five Germans taken by the British to assist in Operation Backfire, a British program to evaluate the V-2, interrogate related German personnel, and launch recovered rockets across the North Sea. None of the British officers told the Germans where they were going and what they were doing, so it was with some trepidation that eighty-four of them left Garmisch-Partenkirchen in a convoy of six army trucks. The Germans were split into two groups, each serving as a means to check facts against the other, with the exception of Dornberger. The program's former leader was isolated for fear that he would convince his men not to work with the British and ultimately sabotage their V-2 program.

By August, the British had amassed all the equipment needed for Operation Backfire, including a nearly complete set of V-2 manufacturing drawings. The assembly and checkout phase was getting underway in anticipation of the program's launch phase that would see rockets flying for data-gathering purposes. But Dornberger was no longer involved. He had been sent to London, ostensibly for further interrogation, but in reality he was taken in as a prisoner of war at a camp for high-ranking German officials. Wearing his camp-issued light brown uniform with no insignia save "PW" emblazoned in white letters on the back of his tunic, Dornberger was then sent to London Cage, where Scotland was waiting. Because Obergruppenführer Hans Kammler was nowhere to be found, Scotland told Dornberger he would be tried in his stead for the crime of launching rockets against England. Dornberger protested, arguing that he had had nothing to do with the German Army's decision to launch the V-2s, that he was only a scientist and hadn't participated in the decisions to deploy the V-2 as a weapon. But Dornberger's arguments fell on Scotland's deaf ears. The scientist's fate would be decided by the cabinet and the chief British prosecutor. With a trial looming before him, Dornberger was eventually transferred to the Bridgend Prisoner of War camp in South Wales where he spent two years theoretically contemplating the atrocities he had committed against Britain.

In 1947, Dornberger was released and managed to emigrate to the United States. He served the U.S. Air Force as an adviser on guided missile development before moving into industry for a job with Bell Aircraft in 1950. Working with a peacetime army and a contractor whose interests went beyond building weapons, Dornberger rehashed the idea of the antipodal bomber, the system Eugen Sänger and Irene Bredt had failed to develop during or after the war. He'd gotten his hands on one of the couple's reports about the rocket drive for a long-range bomber, and now, in peacetime, Dornberger proposed the concept as both a sophisticated weapons system and a research aircraft that could reach speeds of up to six thousand feet per second at altitudes between fifty and seventy-five miles. He twice tried to use the prospect of an antipodal bomber-turned-spacecraft program to court Wernher von Braun into leaving the U.S. Army to join him at Bell Aircraft. Bell, Dornberger alluded to his former colleague, could be the company that would build America's first spaceplane for the U.S. Air Force. Tempted to be on the ground floor of America's first steps into space, von Braun spent sleepless nights turning the decision over in his mind before ultimately deciding to stay the course with the U.S. Army. His team and the Redstone rocket they were building, he reasoned, had an equal chance of launching satellites into orbit. Besides, he felt a strong sense of responsibility for the group at the Redstone Arsenal and an unrelenting pull to develop the rockets and missiles he'd always wanted to build.

Von Braun may not have been convinced, but Dornberger found a sympathetic and willing collaborator in Robert Woods, Bell's chief engineer. In a memo to the NACA dated January 8, 1952, Woods proposed a program that would build off the success of the X-1's supersonic flight by delving deeper into the unknown challenges of hypersonic flight at speeds above Mach 5. Accompanying this memo was a letter from Dornberger outlining a detailed plan for Woods's proposed program that included a test flight program into the ionosphere about 370 miles above the Earth. This research aircraft was a liquid fueled rocket plane, heavily inspired by the boost-glide profile of the Sänger antipodal bomber. The time was right, Dornberger said in the letter, for the aviation industry to start looking at a vehicle that could carry men into the upper reaches of the atmosphere and possibly into space. But the German engineer's vision ultimately went farther to a future where rocket-powered gliders would shrink the world.

"Ultra planes" were Dornberger's imagined inevitable commercial spin-off from developing the antipodal bomber, a way to apply the basic principles of guided missiles to commercial aviation. Born in a world before heavier-than-air flight, Dornberger had seen, within his lifetime, bare-bones wood-and-canvas airplanes taking short hops just feet above the ground replaced by sleek fighter jets and experimental aircraft that could fly faster than the speed of sound. Commercial aviation had flourished into a viable business in the same time frame, carrying passengers around the world in luxurious cabins with unparalleled views. Dornberger reasoned that rocket propulsion would follow a similar path, developing into a commercially viable technology over the course of a half century. Nothing would replace propeller planes for short hops like those between cities in the Continental United States, but ultra planes would use rocket-propelled flight to drastically shorten the travel time between major international cities like San Francisco, London, Calcutta in India, and Sydney in Australia.

Dornberger's ultra planes were a two-part vehicle consisting of a passenger-carrying glider mounted on the upper fuselage of a booster. Both vehicles would have multiple pilots who would use their vehicles' flat bottoms and large triangular wings to increase their gliding range. Before launch, the glider would slide into place along rails on the booster's back. When the two vehicles were sitting horizontally like a traditional airplane, it would look like the booster was giving the glider a piggyback ride. Once mated, the booster-glider stack would be flipped on its end so their noses would be pointing skyward, an impressive and imposing black monolith towering above the ground. The upright vehicles would be mounted on a launch platform sitting on rails. Once fueling was complete, the upright ultra plane would travel along the rails from the hangar through massive concrete passageways he called canyons to the large, circular concrete launch area. It was a system reminiscent of the one Dornberger had pioneered at Peenemünde where upright V-2s were transported by rail to their launch platform.

Ultra plane passengers, meanwhile, would arrive at the airport like they would for any other flight, checking in before proceeding to their assigned gate. Only the gate for an ultra plane flight wouldn't be a typical one with a simple stairway allowing passengers to board their plane. Passengers would have to take a bus from the terminal to a point along the canyon leading to the launch area. From there, they would take an elevator twenty feet down into the launch crater walls where a gantry would grant them passage into the glider's cabin. At other levels, similar gantries would allow maintenance crews access to all levels of the booster and the glider for preflight checks.

Passengers would take their assigned seats inside the main cabin area that would be further divided into smaller units. The seats in these cabins wouldn't be fixed like in a traditional airplane. They would be designed to rotate freely like rocking chairs to keep passengers sitting in a familiar upright position throughout the flight, keeping airsickness and disorientation to a minimum. But this was as far as onboard comforts would go. The cost of fuel for a rocket-powered flight would make carrying excess cargo like flight attendants and inflight meals impossible. And besides, the flights would be so short there would hardly be time for a proper meal service.

But the view would more than make up for the lack of inflight amenities. In keeping with the commercial aviation standard of affording all passengers the most spectacular views possible, the glider's cabin would feature small windows fitted with pilot-controlled sunscreens to protect passengers' eyes from the unfiltered sunlight in the thin upper atmosphere. They wouldn't have cocktails or white linen tablecloths, but passengers would be treated to the awesome sights of the blackness of space and the curvature of the Earth stretching out below them.

With both vehicles fueled and loaded and passengers safely seated in the glider with their seat belts fastened, the ultra plane would finish its railway journey into its circular concrete launch pit. Once in position, the glider pilot would rotate the vehicle so its wings faced any oncoming wind, a simple maneuver that would limit excessive turbulence in the first stages of the ultra plane's climb. The booster's five rocket engines would ignite first to deliver 760,000 pounds of thrust, treating passengers to a mighty roar in spite of the craft's structure and fuel tanks absorbing some of the sound waves. And it would be an uncomfortable ride as well. As the ultra plane left the Earth, passengers would have the sensation of getting heavier and heavier in their seats. Feeling one quarter more than their normal body weight just after launch, the g-forces would increase during the ascent until they would feel themselves weighing three times their normal body weight.

However uncomfortable the high g-forces were, they wouldn't last long. Just two minutes after launch, it would be time for staging. The glider pilot would activate a release mechanism, allowing the small vehicle to slide off the rails on the booster's upper fuselage. Momentum would carry the glider higher while the booster's pilots would guide their larger vehicle to a runway landing back at the airport where it would be towed to the hangar, mated with a new glider, and prepared for another flight.

The glider, meanwhile, would continue toward its destination. Once clear of the booster, the pilot would ignite the glider's three rocket engines for three minutes of powered flight, propelling the small passenger vehicle more than 140,000 feet above the Earth at speeds faster than eighty-four hundred miles per hour. Then the engines would cut out. Inside the cabin, passengers would go from feeling as though they weighed three and a half times their normal body weight to feeling just three-quarters of their normal weight, floating ever so slightly in their seats. The total powered portion of the flight from launch to staging through to the glider's engine shutdown would last just four and a half minutes. The remainder of the flight would be a silent, unpowered, gliding descent. Passengers could sit back, relax, and enjoy the sensation of lightness while taking in the stars shining against the blackness of space before watching the curving Earth rush up toward them.

Far too soon for passengers entranced by the view, it would be time for landing. Still unpowered, the glider's pilots would bring the passenger plane down softly on a runway at the destination airport. Now sitting horizontally, passengers would deplane using a familiar rolling staircase brought flush against the fuselage and walk right down onto the tarmac. A shuttle bus would transport them to the airport's terminal where they could catch a connecting flight or begin their vacation.

The first ultra plane flights in Dornberger's imagined future would be, by the sheer cost of the venture, reserved for the wealthy and social elite, but eventually intercontinental travel would be dominated by rocket planes. Airports in select major cities would serve as hubs for ultra planes, effectively shrinking the world for those eager for international travel. And this would be just the beginning. Dornberger expected these commercial suborbital rocket flights would eventually prove to be just the first step in man's departure from the planet. By making the world more accessible, he anticipated, more people would be inclined to start looking beyond the Earth and out into space. A commercial demand for large boosters for ultra plane flights would also force the technology to develop, eventually parlaying into spin-off technologies such as boosters capable of launching probes to other planets.

Though Dornberger saw his ultra planes as a natural progression for aviation once rocket propulsion became commonplace, the flights he imagined would demand significant technological advances beyond existing rocket power. Before any hypersonic passenger planes could fly, questions about supersonic flight needed answers. Aerodynamic heating was foremost among these problems. The friction these ultra planes would experience while gliding hypersonically through the increasingly thickening atmosphere from near space would heat the fuselage to dangerously high temperatures. Some new cooling method or even some new material would have to be developed before an ultra plane could fly. There were unknowns about the vehicle's structure, too. The gliders would be traveling as fast as Mach 20. Less than a decade after Chuck Yeager broke Mach 1, no one was sure how stable an aircraft would be flying twenty times the speed of sound, nor how to design such an aircraft. If Dornberger's vision for the future was going to come to pass, engineers would have to explore hypersonic flight first, and that meant more research aircraft and more advances to the state of the art of aviation.

Toward the end of 1953, almost two years after Woods proposed a hypersonic research program to the NACA accompanied by Dornberger's pitch for an antipodal bomber-type vehicle, no pilot had managed to fly faster than Mach 2. There was no barrier to break at this speed as there was with Mach 1; flying twice the speed of sound was a psychological goal rather than an engineering one, though still an engineering challenge. The X-1 in which Yeager had broken the sound barrier remained the fastest aircraft at Edwards, but it was old technology and growing increasingly stale every day. Variants like the X-1A, however, helped the aircraft retain its starring position. This version was longer than the original X-1, featuring a bubble canopy for better visibility, larger fuel tanks, and a turbo-driven fuel pump that would allow the engine to operate at full power slightly longer. It was also known to be unstable at speeds above Mach 1.8, though Yeager was sure he could push it above Mach 2 and was determined to secure this record as well. But he wasn't the only one. Now a veteran at Edwards with multiple rocket flight under his belt, Scott Crossfield knew he could reach Mach 2 in the Douglas D-558-II Skyrocket, another rocket-powered vehicle owned by the U.S. Navy. It wasn't designed to fly at Mach 2, but Crossfield had spent enough time in the Skyrocket to know he could push it to this speed record, as Yeager had the X-1A. The two brash pilots were locked in a contest, each vying to be the first to fly at twice the speed of sound.

As an NACA test pilot, however, Crossfield wasn't in the business of setting records. His job was to fly engineering test flights while military pilots like Yeager set speed records. If he was going to take the Skyrocket to Mach 2, he would have to take his appeal all the way to the top, to the NACA's director, Hugh Dryden. It was Dryden who had put the speed limit on Crossfield during his checkout flights in the Skyrocket, preventing him from pushing to Mach 2. It would be a navy pilot, Dryden dictated, who would be the one to push the airplane to its speed limits. But Crossfield's determination prevailed, and he decided to take his case to an intermediary.

Oliver "Perk" Perkins was the U.S. Navy's liaison at Edwards who Crossfield hoped would be eager to secure a speed record for the navy. As a former naval aviator himself, Crossfield offered to make an attempt at reaching Mach 2 in the Skyrocket with "U.S. Navy" stamped all over the project. He sold Perkins on the idea, then urged him to lean on Dryden. If the pressure to fly at Mach 2 was coming from the navy and not from a zealous pilot, Crossfield expected Dryden might be more receptive to the idea. This circuitous and potentially professionally disastrous move was a daring one on Crossfield's part, but it paid off. Perkins took the case right to the Pentagon, and a week later Dryden called Walt Williams at the Muroc Flight Test Unit to say that the Skyrocket's speed ban had been lifted, but only for one flight. Crossfield had one shot to break Mach 2. If he missed, it would be Yeager's record to secure in the X-1A. A delighted Crossfield promised Williams that he wouldn't miss.

Friday, November 20, 1953, dawned cold and blustery in the desert. Shivering and weak from a recent bout of the flu, Crossfield arrived at Edwards before daybreak far more concerned with the Skyrocket's health than his own; he was sure he could summon his mental and physical strength for the four minutes of powered flight. He found the Skyrocket nestled underneath the belly of a B-29 launch plane, surrounded by the swirling liquid oxygen vapor that accompanied fueling operations.

Crossfield recognized that he was asking the Skyrocket to perform a small miracle. As such, he resolved to do everything he could to help the vehicle along. He and the ground crews had developed a few tricks to squeeze as much speed out of the aircraft as possible. They figured out that shooting super-chilled liquid oxygen through the engine right before it was launched increased its performance. Crossfield had also learned, through trial and error on previous flights, the best sequence in which to light the engine's barrels to get the most power out of all four combined. But nothing would help push the Skyrocket past Mach 2 more than carrying more fuel, and they had even come up with a means to increase the aircraft's fuel capacity through cold soaking, a process that involved filling the tanks with super cold liquid oxygen hours before launch and letting it settle so the tanks would stretch out and hold a precious few extra pounds of fuel. The launch crews had also perfected a method for topping off the liquid oxygen tank right before releasing the Skyrocket, so Crossfield would have the most fuel available to burn. Crossfield had even had the ground crew wax and polish the Skyrocket's fuselage so it would slice right through the air with the least amount of friction. He had one shot to break Mach 2, and he was pulling out all the stops to get there.

The B-29 took off into morning skies, and after an hour and a half, the bomber with the Skyrocket shackled to its belly was at its launch altitude of thirty-two thousand feet. The mother ship's pilot released the Skyrocket, and Crossfield lit all four rocket barrels in rapid sequence to begin a smooth, shallow ascent into the upper atmosphere. He had calculated that his fuel load would afford him about two hundred seconds of powered flight, and getting the most from every second meant following a very precise arcing parabolic flight path; any deviation could cost him precious speed. The Skyrocket rose to seventy-two thousand feet before Crossfield pitched it over ever so slightly with his rocket engine still blazing. The cold soak had done the trick. The Skyrocket's engine burned for a full 207 seconds before running out of fuel. In the cockpit, Crossfield watched as the Mach meter on his instrument panel edged over the 2.0 mark. He reached a top speed of Mach 2.005, a hair over twice the speed of sound but enough to secure the record.

With all his available fuel consumed, the Skyrocket's engine abruptly shut down, throwing Crossfield forward against his restraints. The now silent and powerless aircraft started losing speed and altitude, gliding toward the ground and easing back into subsonic flight. Just twelve minutes after he'd launched from the B-29, Crossfield brought the Skyrocket to a smooth landing on the Rogers dry lake bed. A press conference at the Statler-Hilton hotel in Los Angeles the next day secured Crossfield's place in history with reporters jockeying to interview the fastest man alive.

Twenty-two days later, Yeager made his own attempt to break Mach 2. Having failed to secure the record first, he at least wanted to break Crossfield's record and take back the title of fastest man alive right in time for the fiftieth anniversary of the Wright brothers' first flight on December 17. The first time Yeager had taken the X-1A out earlier that year, it felt familiar, and the two flights that followed had been equally smooth. His fourth flight, on December 12, was different, however.

The day started normally enough for Yeager, with a couple of hours spent hunting before he arrived at Edwards. After a light breakfast and a forced delay to mend a minor problem with his pressure suit, time he used to clean his shotgun, the X-1A was mounted underneath its B-50 launch plane, and Yeager was ready to go. At thirteen thousand feet, he climbed through the bomb bay into the small cockpit and checked out his systems before giving the B-50 crew the all clear to lower the domed canopy over his head and bolt it into place. Imprisoned at thirty thousand feet, Yeager heard the familiar sound of the shackles releasing the rocket plane from its mother ship and felt himself rise up in his seat at the sudden fall. He reached for his ignition switch, lit three rocket barrels, and watched as shock waves danced over his wings. With his nose slightly higher than anticipated, Yeager hit eighty thousand feet, and just before his engine ran out of fuel he hit Crossfield's Mach 2.005. Then he pitched the unpowered X-1A over to gain more speed and watched as the needle on his Mach meter rose to 2.4.

It was only then that Yeager realized he was flying too fast at too high an altitude. One of his wings kept coming up, forcing the aircraft to roll over, and in the thin upper atmosphere he couldn't fight the rolling motion. Bell Aircraft's engineers had warned him not to take the aircraft above Mach 2.3, and they had been right. The roll had started right as Yeager hit Mach 2.4, then the X-1A started tumbling as it fell from the sky. Yeager was thrown around inside the cockpit so forcefully that his helmeted head broke through the canopy and nearly caused him to lose consciousness. Sensing a loss of pressure in the cockpit his pressure suit inflated, which immediately fogged up his faceplate. Yeager had half-obscured glimpses of light and dark, the Sun and the ground alternatively flashing by his line of sight as he continued to tumble. All the while, he could do little more than mumble unintelligibly over the radio. Still blind from the fogged helmet, Yeager groped over the familiar instrument panel and found the switch to readjust his rear stabilizer. It did the trick, and in the thicker atmosphere at thirty thousand feet the X-1A entered into a normal spin, something Yeager knew how to get out of. He recovered in just five thousand feet. Dazed and unsure whether the aircraft was too damaged to fly, Yeager managed to bring the X-1A to a safe landing on the dry lake bed at Edwards Air Force Base. Yeager's skill and a fair bit of luck had saved his skin on what could have been a fatal flight.

Crossfield's flight in the Skyrocket and Yeager's near-death experience in the X-1A underscored a serious discrepancy. It was clear that aviation as an industry needed a new research plane to address the problems of flights in excess of Mach 2 if anything was eventually going to fly higher and faster. There was one aircraft under development that promised to make great strides, Bell Aircraft's X-2. The X-2 was designed to expand the speed and altitude regimes of the X-1, but the program had been continually stunted by development problems, leaving it languishing in a hangar when it was needed in the sky. Desperate to see this powerful plane fly, Crossfield had asked Hugh Dryden if he could be loaned to Bell on a special assignment just to get the X-2 flight ready, then return to Edwards with the new research plane in tow. However useful it might have been to have an NACA representative pushing for the X-2's completion, Dryden rejected the proposal. Crossfield was, Dryden countered, needed at Edwards.

Luckily for Crossfield's desire for a more advanced research vehicle, Woods's 1952 memorandum pitching a hypersonic aircraft had eventually found some support at the NACA. The NACA Committee on Aeronautics wasn't immediately interested in pursuing Woods's proposed program but didn't disregard the idea right away either. The proposal was shelved until a meeting that June during which the committee passed a two-stage resolution. First, it recommended that the NACA undertake a research program into flights up to fifty miles at hypersonic speeds between Mach 4 and Mach 10. The second called for some future program that would deal with flights above fifty miles at speeds from Mach 10 to escape velocity, speeds fast enough for the vehicle to achieve orbit. Spaceflight, however poorly understood and seemingly futuristic, was already in the minds of key decision makers. These early inclinations toward spaceflight spawned some early proposals, one of which called for a supersonic mother ship to launch a rocket plane fitted with Sergeant rockets developed at Caltech's Jet Propulsion Laboratory to extremely high altitudes.

It took two years for the proposed hypersonic research program to take the first steps from concept toward reality. By 1954, experts agreed that the potential of rocket-powered aircraft, and particularly hypersonic flight, was exciting, but they also recognized that the future of hypersonic rocket-powered flight hinged on major advances in all areas of aircraft design.

Sänger's 1944 assumption that a boost-glide vehicle would demand only minor new technologies proved to be erroneous. Aerodynamic heating was one known problem, the so-called thermal barrier. Another was the challenge of flying in the thin upper atmosphere. Airplanes have control surfaces, ailerons, rudders, and elevators that push against the air to move the vehicle. But at altitudes where the air is too thin for traditional flight controls to push against, a pilot would need some other means of control. Hypersonic flight would remain fodder for science fiction until engineers could devise a high-altitude flight control system into a vehicle that could survive a punishingly hot reentry profile.

Luckily, the time was right for such an aircraft. Coming on the heels of the Mach 2 flights, consensus in the aviation industry was largely in favor of a continued rapid increase in speeds. That the engineering problems couldn't be solved in ground testing was another factor pushing for a hypersonic flight research program. A new research airplane would be the test object and the sky, its laboratory. It would be powered by the same rocket engines that were powering the missiles being developed by the armed services. That there was no competing program under development helped push the NACA's hypersonic research aircraft forward. After early successes, there was ample political and industry support for the X-series of aircraft to continue. The U.S. Air Force Science Advisory Board Aircraft Panel also believed that the time was right for a new NACA-military program and stood firmly behind the idea of developing a research aircraft to gather data at Mach numbers from 5 to 7 at altitudes of several hundred thousand feet.

The bureaucratic necessities to bring this hypersonic program to life were also falling into place in 1954. Dryden, a lifelong proponent of super- and hypersonic flight research and the NACA's director, was named chairman of a new Air-Force-Navy-NACA Research Airplane Committee. This committee was dedicated to collecting experimental research aircraft data, exploring the problems of piloted flight at high speeds and altitudes, as well as guiding the development of an airplane to explore the problems of flight at the highest speeds and altitudes possible. The proposed hypersonic aircraft met the committee's needs and fell under a now-familiar arrangement: the military would fund the development and construction of the aircraft that the NACA would use in its flight research program, and both the military and the civilian agency would benefit from the research results.

A hypersonic research program was formally initiated in February 1954 with goals ranging from probing these new flight areas in order to gather data to developing operational supersonic fighter aircraft that could fly between Mach 2 and Mach 3. Early studies set the basic design constraints such that the aircraft would address major areas of interest. Engineers with the NACA determined that the best way to return from high altitudes was to have the aircraft's nose aimed toward the sky, a high angle of attack configuration that would expose the aircraft's whole underside to the atmosphere, acting as a large aerodynamic brake. Preliminary studies revealed that control and stability were problems in desperate need of an answer; no one wanted a repeat of the flight that had nearly killed Yeager.

The launch configuration would also have to be different. The B-29s were being phased out and the next logical choice, the B-36, was too unknown to the team at the High Speed Flight Station and Edwards Air Force Base for them to turn it into a viable launch plane. The larger B-52 emerged as the best option for a new mother ship, but it lacked the large central bomb bay of the B-29. The hypersonic plane would have to be launched from underneath one wing. This practical decision introduced more unknowns, namely how the mother ship would take off and fly with an asymmetrical load and offset center of gravity that would change in flight at the moment of launch. Another consideration was the flight path of this proposed aircraft. To this point, all rocket aircraft had launched in the skies over and landed at the Rogers dry lake bed. Their powered flights were short and relatively low, allowing them to be monitored from ground stations at

Edwards. This wouldn't work for the hypersonic aircraft designed to fly higher and faster than anything else. If it were to land on Rogers, it would have to launch over another lake, and the pilot would have to rely on other dry lake beds in the vicinity if he ran into some midair emergency. Extending the physical space of the flight in turn demanded a better communications and tracking system.

By July, the NACA had completed its studies and was ready to present its hypersonic aircraft, designated X-15, to the U.S. Air Force and Navy. The preliminary concept set the basic design requirements of an aircraft obviously heavily influenced by missiles. Forty-eight feet long with stubby wings just twenty-seven feet across in the middle of the fuselage, it was mainly the small bump with two narrow windows toward the nose of the fuselage that made it clear that the vehicle was a manned aircraft and not a pilotless missile. A thick, wedge-shaped vertical stabilizer featured prominently in the aircraft's rear, something engineers found broke up airflow to eliminate the kind of instability that had nearly killed Yeager. One Langley researcher had added a split trailing edge on the vertical stabilizer, something that could act as an additional speed break.

The NACA design also featured an x-shaped empennage, the stabilizing surface at the tail end of the aircraft, to bring increased stability and control to the high-speed flight through the thin upper atmosphere. To minimize the impact of aerodynamic heating, the NACA study specified that the X-15 be made of Inconel X, a nickel alloy that could withstand the intense temperatures the aircraft would be subjected to during reentry. Even though it wouldn't be going into orbit, the X-15 would still descend quickly through multiple layers of the Earth's atmosphere. To reach its prescribed high altitudes, the aircraft would have to fly a ballistic profile not unlike a missile, arcing high into the upper atmosphere before curving back down again. And the height of this ballistic path meant the aircraft would reach air thin enough to demand a system of hydrogen-peroxide-powered reaction controls for attitude control. The hypersonic aircraft emerged from this study as a futuristic vehicle, but for all the technological advances wrapped up in the program space wasn't overtly in the cards. High-altitude flight was one thing, but *space* was still a dirty word as far as the air force was concerned. The service was in the business of expertly engineered, technologically advanced aircraft, not simple vehicles designed to leave the atmosphere.

Langley released the specifications of this conceptual research aircraft during a meeting at the NACA headquarters on July 9, 1954. In attendance were representatives from the aviation industry, the NACA, the air force, and the U.S. Navy by Dryden's invitation; he wanted to get all military branches involved as well as the Department of Defense, though the hypersonic aircraft would be primarily a joint air force–NACA program with navy support. When the committee met again in October, Scott Crossfield was on hand as one of the NACA representatives. Having reviewed historical data relating to the project, he outlined for attendees the performance requirements for this new aircraft. And after watching the X-2 languish and coming face to face with the challenge of flying at just Mach 2, Crossfield was desperate to see this hypersonic research aircraft brought to life on schedule. But there were development challenges facing the project that went beyond the immediate technical aspects. Largely thanks to the Korean War, missiles were advancing in leaps and bounds and were routinely flying faster than Mach 10. This meant the technical side of hypersonic research was understood, leaving the human factor as the largest unknown. And this new program promised to take a massive leap. Instead of a series of vehicles designed to fly incrementally faster, this new aircraft was going to jump from Mach 2 to Mach 7, more than tripling existing speed records in one fell swoop. It was an audacious goal that risked becoming contro-versial enough to kill the project.

But once it gained momentum the X-15 program moved quickly. In December, the air force's Air Materiel Command invited prospective bidders to submit their proposals. Four were received for evaluation by the NACA the following May from Bell Aircraft, Douglas Aircraft, North American

Aviation, and Republic Aviation. Each proposal had its merits, and each contractor brought different experience to the program. Bell Aircraft, arguably the most experienced company when it came to building successful rocket-powered aircraft, pitched an airplane as simple and clean as the X-1. Douglas Aircraft also drew on its past successes, presenting a hypersonic aircraft reminiscent of the Skyrocket. Republic Aviation was at the time working on a Mach 3 interceptor aircraft and brought its relevant research to its X-15 proposal. Only North American Aviation had no experience building supersonic, rocket-powered aircraft, but it did have some experience building missiles; it was working on a winged cruise missile called Navajo that used the V-2 as its jumping-off point.

To Crossfield, North American was the default winner right off the bat. Bell Aircraft was fast falling out of favor with the NACA over the X-2 debacle. Douglas Aircraft's proposal was sound but had shirked the NACA's recommendation to use Inconel X in favor of a material called HK31, which was significantly thicker, heavier, and ill-suited to the X-15's demanding flight profile. Republic Aviation almost seemed to misinterpret the NACA's guidelines in its proposal, pitching an aircraft that emphasized high speed over high altitude. North American Aviation's proposal was the most straightforward, essentially giving the NACA its concept aircraft right back with just minor changes. North American's version of the proposed X-15 was slightly longer at fifty feet with a slightly narrower twenty-two-foot wingspan. The NACA's High Speed Flight Center, Ames Research Center, and Langley Research Center each evaluated the four proposals. Ames and Langley preferred North American's proposal, while the High Speed Flight Center alone preferred Douglas's. But its strong standing in the aviation world certainly helped push North American over the top. The company's P-51 Mustang fighter planes were among pilots' favorites of the Second World War and its F-86 Sabres were among the earliest American swept wing supersonic fighter jets taking on Soviet MiG-15s in dogfights over Korea. The final standing in July saw North American emerge as the winner. Its proposed aircraft wasn't perfect, but it was on the right track. And to the chagrin of the military footing the bill, the winning bid came with the highest price tag of $56.1 million.

After submitting its proposal, however, North American turned out not to be too keen on the idea of building a small run of three highly experimental aircraft, and the company requested to withdraw its proposal in September. Hugh Dryden was far more inclined to reopen the bidding process rather than award the contract to Douglas, the runner-up contractor, though the prospect wasn't appealing. Hoping that North American would eventually change its mind, Dryden opted to continue with the procurement process in spite of the contractor's reservations, though it didn't look promising. North American was busy with other, much larger projects and remained unwilling to divert manpower to a small research project. The NACA offered to extend the proposed production schedule by eight months, but still North American's chief engineer, Raymond Rice, shied away from accepting the contract.

One man, however, was extremely anxious to secure the X-15 contract for North American. Harrison Storms was the manager of research and development at North American's Los Angeles division. After studying under Theodore von Kármán at Caltech, Storms had cut his teeth in the aviation industry with North American during the Second World War. He had joined the company right after the Japanese attack on Pearl Harbor and devoted himself to building better airplanes for the American Army Air Force. He had been with the company ever since. One day in the midst of the X-15 contract discussions, Rice called Storms into his office. He could have the X-15, Rice told Storms, on one condition. Storms would have to take full charge of the program, acting as the top North American representative on the project, and keep every single problem off of Rice's desk. With this new arrangement, North American finally accepted that it had won the X-15 contract. The first hypersonic research plane thus fell under a team led by Storms and Charles Feltz as chief project engineer.

Storms might have been eager to get the X-15 program started, but Scott Crossfield had some lingering concerns. He worried that something as conceptually big as a hypersonic research aircraft would stall in the development stage and become stagnant like the X-2 had done for so long. So he went to Walt Williams at the High Speed Flight Station and again made a very specific and very unorthodox request of his director: he wanted to be officially assigned as the NACA liaison at North American Aviation on the X-15 project. He wanted to be on hand with the contractor to ensure the program met the NACA's constraints and stayed on schedule. But Williams denied Crossfield's request, again making the argument that Crossfield was needed at Edwards. Crossfield, however, disagreed, and was far firmer in his conviction this time around. When Williams refused to give in, Crossfield tendered his resignation from the NACA.

His unemployment pending, Crossfield traveled alone to North American Aviation's Los Angeles division, the site where the X-15 would be built. Luckily for Crossfield, having been the first man to reach Mach 2 meant he was known and respected in the aviation community, and he had already met a handful of industry giants, among them the president of North American Aviation, Lee Atwood. Crossfield went to Atwood's office and presented the same proposal he had to Williams. He told Atwood he was a man who could bring a valuable wealth of knowledge, experience, and familiarity with rocket planes to an organization that was taking on this exotic challenge for the first time. He wanted to be part of the program from the start, from the initial designs all the way through the airplane's entire construction and flight test program, making the first test flights before the X-15 was handed over to the NACA and U.S. Air Force for the record-breaking flights. It was an unusual proposal for Atwood, but both he and Rice signed off on it.

Crossfield left North American's Los Angeles division having created a job for himself, but he remained with the

Muroc Flight Test Unit for a few months. The Muroc Flight Test Unit had changed drastically in the five years since he first toured the desert site. Originally run by the NACA's Langley Memorial Laboratory, Congress had allocated funding to transform the satellite site into a new outpost in 1951. On July 1, 1954, the Muroc Flight Test Unit had been rededicated as the High Speed Flight Station, the newest NACA site with strong in-house capabilities to demonstrate the advances in high-speed aeronautics. Crossfield's imminent departure, however, opened a spot for a new test pilot to join the NACA's contingent at Edwards Air Force Base. To Neil Armstrong, also a young pilot engineer with a passion for research, Edwards was mecca.

CHAPTER NINE

Edging into Hypersonics

In December 1955, Scott Crossfield pulled his car into a parking lot near a group of large manufacturing buildings on the south side of Los Angeles International Airport. Together, the buildings housed North American Aviation. Crossfield made his way to Building number 20, a Second World War–era relic that now housed the company's cafeteria. Alongside the cafeteria was a small, cramped space employees called the garret. A sign on its door read SECRET, UNAUTHORIZED PERSONNEL PROHIBITED. Though right next to where they ate meals, no North American employee could pass through the door without being cleared to enter the area and signing into a logbook. Even then, every visitor needed an escort at all times. Crossfield, however, was cleared to enter the cramped space on his own. Inside he found a series of desks placed too close together where nearly a dozen men sat poring over papers and technical drawings. He took his place among the small team assigned to turn the X-15 from concept to reality. After years of dreaming of a hypersonic research aircraft to soaring through the skies over Edwards, Crossfield could finally start thinking about this plane in terms of real flight hardware.

Everything about the X-15 was new and different. A giant in the aviation industry, North American usually built planes the way major auto manufacturers built cars in Detroit. Once an aircraft's design was agreed upon by the engineers in charge, it was frozen, and the company set out to build a large run of its newest product with the efficiency of an assembly line. But the X-15 was different. Not only was it far more complex than any of North American's previous aircraft, it had a very small production run with just three units contracted by the NACA. And because it was a specialized program, management was reluctant to take resources away from the larger programs that promised a financial return. As such, the X-15 became something of an anomaly, a project run by a small special team under the company's Advanced Design Section, wholly independent from every other part of North American.

After arguing with Raymond Rice to allow North American to take on the X-15 project, Harrison Storms was the overall manager of the program. But management of all day-to-day operations was assigned to Charlie Feltz, the chief project engineer. A veteran mechanical engineer whose career began around the start of the Second World War, Feltz had never heard of the hypersonic research plane until the assignment landed on his desk and he became the leader of a skeleton crew of ten men. Crossfield's arrival increased Feltz's team to eleven, and the pilot was almost as much of an anomaly to Feltz as the X-15 was. The pilot turned consultant didn't work directly for Feltz, he hadn't been hired by Feltz, and his duties were vaguely undefined. Crossfield was there to lend his experience as a supersonic pilot to the team, a role that Feltz finally called Design Specialist. It was a somewhat haphazard title that suited Crossfield's role as technical adviser and expert as well as anything.

As Feltz's team started hashing out the details of the X-15 based on the original proposal, it revealed itself as an increasingly complicated machine. Externally it was simple and sleek, a fairly traditional design for a rocket-powered aircraft. Its tall, thick vertical tail, elongated nose, and smooth body with a V-shaped canopy barely sticking up above the fuselage were fairly standard, as were its short, stubby wings jutting out from both sides. The most obviously unconventional thing about the X-15 was its power plant.

The early concept drawings listed the engine as an XLR-99, a throttleable liquid-fueled rocket engine capable of delivering fifty-seven thousand pounds of thrust at forty thousand feet, meaning it could match most missiles in terms of its power output. Equivalent to about one million horsepower, the XLR-99 engine promised to accelerate the small plane to speeds in excess of 1.5 miles per second or nearly 8.7 million miles per hour, equivalent to about Mach 7 at altitudes over 250,000 feet. The X-15 would more than double existing speed and altitude records.

Crossfield realized before too long that once the X-15 started making regular flights at breakneck speeds to unprecedented altitudes, engineers and scientists nationwide would be clamoring to add their experiments to the aircraft. But every addition threatened to add weight and delay production, so he gave himself an unofficial role as "the X-15's chief son-of-a-bitch." Nothing on the aircraft would change without going through him first. He was going to keep it on schedule come hell or high water.

As soon as they started digging into the intricacies of the X-15, Feltz's team unveiled no shortage of wrinkles that needed to be ironed out.

Managing the X-15's flight profile was one problem. It was clear that with a peak altitude above 250,000 feet the X-15 would need some kind of reaction control system to keep the aircraft oriented in the upper atmosphere; short bursts of compressed gas would allow the pilot to trim his attitude, adjusting his aircraft's orientation in flight. But the pilot would also have to transition seamlessly from reaction controls to traditional flight controls on his long, gliding descent. Some kind of testing and development of this new system was in order.

Atmospheric heating was another issue. Though the X-15 wouldn't get as hot as Walter Dornberger's proposed ultra planes, it would still be flying in a heat region about which fairly little was known. Construction materials would help; Inconel X, the steel-nickel alloy, was one strong ally in the X-15's fight against high heat. But the edges of the aircraft that would bear the brunt of the heat would need something more to protect them. The initial solution was to add an ablative coating to these edges, something that could harmlessly burn away to protect the vehicle from the hottest portions of the flight. But this coating not only added weight to the aircraft, wind tunnel tests revealed that ablative leading edges would make the X-15 aerodynamically unstable. The only solution was to make the leading edges solid Inconel X, an effective but heavy solution.

In the first months after Crossfield's arrival at North American, the X-15's development crew grew steadily, but so did the aircraft. Every pound added meant the aircraft would fly slightly slower and reach a slightly lower peak altitude. The use of solid Inconel X was only one matter. Another was the NACA's stipulation that the X-15 have 3 percent fuel ullage allowance to account for the fact that the tanks could never be completely filled. But 3 percent of the aircraft's eight tons of fuel translated to a loss of about two seconds of flight, a seemingly small but quite significant performance penalty. The solution was to increase the aircraft's diameter to allow for larger fuel tanks, increasing its fuel capacity by twenty-five hundred pounds. But this solution also added weight. Small adjustments like this added up until the plane weighed in at thirty-one thousand pounds, at which point Feltz had to draw the line. The X-15 couldn't get any bigger, he told his team, and urged the men to shave off as many ounces as they could from any nook or cranny that could spare it.

The issue of weight became more complicated months into the X-15's development when updated information on the XLR-99 engine said its power output would be lower than initially anticipated. There was no way to shave off enough weight to account for the now less powerful engine, but Feltz came up with an elegant solution to increase the aircraft's lift to regain the lost speed and power. From the start, X-planes had maintenance tunnels running along the top and bottom of their fuselages. These were large, pipelike housings through which wires, control cables, and plumbing tubes were routed; because the fuel and oxidizer tanks took up most of the space inside these aircraft, routing plumbing and power cables around the tanks was a necessity. Feltz wondered whether the tunnels could be moved to the sides of the fuselage, broadening the base to increase its lift. Wind tunnel tests revealed his instinct was a good one. The side-mounted tunnels increased lift, and cutting them off just behind the cockpit combatted the strange aerodynamic phenomenon observed in wind tunnel testing that caused the nose to pitch up in flight.

Questions also swirled around the X-15's landing. Like the rocket planes that came before it, the X-15 was designed to glide to an unpowered landing on the dry lake bed at Edwards Air Force Base, touching down on rear skids and a forward nose wheel. But the tail became a problem. Early design studies and wind tunnel tests said that the best aerodynamic shape for the X-15's vertical tail was a diamond shape as seen from an overhead perspective. Extending that diamond-shaped tail above and below the fuselage would bring the same stability to the high speed portions of the flight. But not only was it a heavy design, this tail extended beyond the skids and landing gear. The team half-joked that the tail promised to turn the X-15 into the world's fastest plow upon landing.

The whole X-15 team pored over drawings and blueprints together trying to find a solution to the plow problem. They finally brought Storms into the mix, the absent leader whose responsibility for the program's success meant he was never more than a phone call away. In instances like this when he was called to bring his expertise to a problem, Storms descended on the team in a manner befitting his last name. He considered the X-15's tail and offered what he viewed as an obvious solution: use an explosive charge to detach the lower portion of the tail before landing. It was only needed during the high-speed portions of the flight, so why not get rid of the lower portion once it was no longer useful? Storms similarly offered an elegantly simple way to shave weight off the aircraft. The forward half of the diamond tail was necessary, but the rear half was little more than the completion of the shape. Cutting it in half, turning it from a diamond into a wedge, could literally halve the weight of the tail. Wind tunnel tests confirmed that Storms's instincts were right, and the changes were made.

But for every problem the team solved, another one soon took its place. At one point in 1956, the U.S. Air Force alerted North American to a new ruling that said every air force plane had to have an escape pod for the pilot in lieu of a traditional ejection seat. Not only would an escape pod add about twelve hundred pounds to the X-15 at the cost of about one Mach number in speed, the related pyrotechnic system would further complicate the aircraft. Not only that, but Crossfield was dead set against an escape pod. The Douglas Skyrocket in which he'd reached Mach 2 had had an escape pod system that Crossfield swore he would never use after early testing said the g-force associated with a pyrotechnic escape would almost certainly be fatal.

Feltz agreed with Crossfield. In an emergency, the X-15's cockpit was probably one of the safest places in the world. It was pressurized with non-flammable nitrogen gas so there was no risk of a fire, and the whole thing was reinforced to protect the pilot against the high g-forces of a hard landing. It would be far safer for a pilot to remain in the X-15 and eject at a slower, safer speed. Not to mention that making this kind of addition could only make the overall project more expensive and delay the X-15's initial flight by as much as a year. Changing the air force's mind, however, wasn't as easy as calling the service to ask for an exception to the escape pod rule. So Feltz's team was forced to go through engineering the escape pod meticulously to show exactly why it wasn't necessary for the X-15.

When the first cockpit mockup was completed in July 1956, it included a specially designed ejection seat with small stabilizers on the sides that would limit any oscillations upon ejection. The pilot would land by personal parachute. When air force representatives visited the North American facilities, they listened to Crossfield's briefing and said nothing about his exclusion of the ejection pod. Crossfield, it seemed, swayed the air force men to his way of thinking. They signed off on the design, which passed with flying colors. It was one instance, Crossfield felt, when he truly lived up to his self-proclaimed title of the X-15's chief SOB.

Feltz's X-15 team had to deal with more than just technical flight considerations. The pilot's comfort was another matter. Unlike what Chuck Yeager had done with the X-1, the X-15 pilot couldn't ride up to altitude in the mother ship and climb into the rocket plane's cockpit when it was time to launch. The launch plane for the X-15 had changed from the piston-powered B-36 to the jet-powered B-52, a larger bomber

Concept art from 1961 showing the Dyna-Soar launching atop a Titan II missile.

Above left: John Paul Stapp riding a rocket sled at Edwards Air Force Base.

Above right: The X-15 nestled under the wing of its B-52 launch plane.

Right: Pilot Stan Butchart training in the Iron Cross Attitude Simulator at the NACA High-Speed Flight Station.

Above: Neil Armstrong training in the Iron Cross Attitude Simulator at the NACA High-Speed Flight Station in 1956.

Above: Scott Crossfield standing in front of the Douglas D-558-II Skyrocket in which he broke Mach 2 on November 20, 1953.

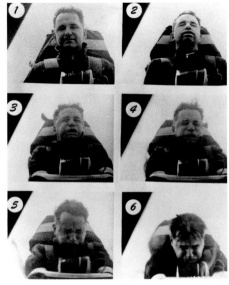

Above: Six frames showing the effects of a rocket sled deceleration test on John Paul Stapp.

Right: The V-2 rocket, tested and developed by the German Army, is readied for launch under the British Operation Backfire after the Second World War.

Below: An American soldier examines a half-completed V-2 rocket in an underground assembly plant after the Second World War.

Above: Fritz von Opel after driving the rocket-powered car Opel RAK 2 at the Avus Speedway in Berlin.

Left: Wernher von Braun speaks with Harrison Storms during a 1960s visit to the North American Aviation's Space and Information Systems Division in Downey, California.

Below: NACA High-Speed Flight Research Station groundbreaking ceremony on January 27, 1953.

Above: The rear end of the rocket-powered car Opel RAK 2 at the Avus speedway in Berlin, just before Fritz von Opel took it for a test drive.

Right: American Air Force Lieutenant Colonel David G. Simons, Otto Winzen, and Vera Winzen with the Project Manhigh gondola in 1957.

Below: President Eisenhower with NASA Deputy Administrator Hugh Dryden, left, and NASA Administrator T. Keith Glennan, right, in 1958.

Above: The opposite of a streamlined space plane, a full-scale Mercury blunt-body capsule goes through wind tunnel testing in 1959.

Above: JPL director William H. Pickering, physicist James Van Allen, and Wernher von Braun holding a model of Explorer 1 after it successfully reached orbit.

The US Army's Jupiter C rocket launches America's first satellite, Explorer 1, into orbit on January 31, 1958.

that would launch the X-15 from a pylon mounted underneath one of its wings. This meant the rocket aircraft's pilot would be in the cramped cockpit for hours, waiting on the ground during the final fueling and checkout and throughout the B-52's ascent. He would need some protection against stiffness and pain from the vibrations of the mother ship, but that protection had to be lightweight. Even a foam rubber cushion could add a devastating two pounds to the aircraft's overall weight.

The team in the garret sat down and asked themselves what company in the country had the most experience keeping men comfortable in a seat in rough conditions for hours at a time and came up with the International Harvester Company, a purveyor of tractors. Part of the company's success was due to its investigation of the human spine's natural frequency—its response to sitting in a vibrating vehicle for a length of time—and subsequent design of a seat that could protect it while driving over hard terrain. The seat in the X-15 was designed as an exact duplicate of a tractor seat.

But honing the technical design of the X-15 was only half the battle. The pilot needed to be more than just comfortable, he had to be kept alive at extremely high altitudes. At sea level, atmospheric pressure is high enough that humans don't have to wear protective garments. But above forty-five thousand feet, the atmosphere is so thin that blood wants to escape into the less dense environment and skin can't compress the body enough to stop this from happening. With the X-15, flying higher than 250,000 feet, the pilot would need some protection against the thin atmosphere. The solution was a pressure suit. Pressure suits do at altitude what skin can do at sea level, compressing the body and exerting a restraining pressure on the skin to mimic the safety of sea level. In this case, the suit would have to be self-cooling to protect the pilot from succumbing to heat exhaustion and also provide him with enough oxygen to survive the flight. It would also need some arrangement of rubber bladders that could inflate at the first sign of high g-forces to protect the body during a high-speed ejection.

Opinions were divided on the best pressure suits. At the time the U.S. Air Force favored a partial pressure suit, a cloth suit that simply squeezed the body with enough force to offset the effects of low atmospheric pressure. The navy preferred a full-pressure suit, a self-contained unit that could have applications far beyond high-altitude flights. The right full-pressure suit could protect men from the vacuum while walking on the surface of the Moon, which some forward-thinking navy engineers expected to see happen sooner rather than later. With this futuristic application in mind, David Clark had designed a full-pressure suit for the U. S. Navy that was like a wearable inner tube. An inner layer of rubberized nylon would inflate at altitude to compress the body and the outer layer of cloth would keep the nylon firmly in place. It was a flexible design that wouldn't constrict the pilot's movements much, thanks to the fabric. In the 1930s, David Clark had invented a knitting machine that could sew a seamless piece of fabric with two-way stretch. It became extremely popular for girdles and brassieres and allowed him to corner the market. His company had expanded to military garments beginning with the Second World War, but women's under-garments remained his bread and butter.

Crossfield strongly preferred the navy's Clark-designed full pressure suit for the X-15. He had first discovered it in 1951 when he visited a navy laboratory in Philadelphia after his assignment to the Skyrocket program. He had donned the suit, climbed into an altitude chamber, and waited patiently as mechanics removed enough air to simulate an altitude of ninety thousand feet. A pleased Crossfield was somewhat stunned to learn that his test was the first time the suit had been used at such a simulated high altitude. Its benefits widely known, a version of the Clark suit had been included in North American's bid for the X-15, and under Crossfield's supervision it was on track to be the official life support system for the rocket plane's future pilots.

Over the course of his first year with North American, Feltz's team was still feeling its way to bringing the X-15 to life. Crossfield remained the most familiar with rocket planes, but the group's spirit made up for their collective lack of experience. The last half century had been a battle against drag and gravity, slowly developing the streamlined and efficient machines that would fight gravity and carry men off the ground. Now, this one aircraft—promising to send men into a wholly unknown region of near space—was coming together in record time, and the transition from final design to actual flight hardware couldn't come soon enough.

With the X-planes at Edwards growing increasingly antiquated, the need for the new research aircraft was fast becoming pressing. Coming on the heels of the X-1 as its immediate predecessor, the X-2 was intended not only to fly higher and faster than anything that had come before it but to withstand higher temperatures as well. This aircraft had been designed to begin tackling the problem of aerodynamic heating, clearing the way for the X-15 and later vehicles to push toward space. Made of stainless steel and a high-strength copper-nickel alloy called K-Monel, the aircraft's fuselage was prepared for the heat associated with a Mach 3 flight.

Many problems had dogged the X-2. Proposed not long after the Second World War ended, this aircraft's high speed, high altitude, high heat flight profile hinged on the development of a host of new technologies, not all of which made it into the final aircraft. One unrealized system was fly-by-wire, a system that would feed the pilot's control inputs from the cockpit into a computer that would in turn operate the aircraft's control surfaces with motors. It was a decision that came down to scheduling. Though easier to fly and acting as a failsafe by checking the pilot's inputs, fly-by-wire was eventually abandoned in favor of a conventional hydraulic system that could be incorporated into the aircraft sooner. This heavier system that had the pilot's control inputs directly move the aircraft's control surfaces left control squarely in the hands of the man in the cockpit. The X-2 also used a conventional gyroscope to give the pilot information on his orientation, a device typically so inaccurate at high altitudes that it became basically unusable.

Shortcomings aside, when the X-2 finally had rolled out and made its first glide flight in 1952, it was a very welcome addition to the hangars at Edwards where the X-1 was entering its fifth year as the most advanced aircraft on site. Now four years later, in 1956, the X-2 had racked up a mixed bag of successful missions and failed or aborted flights under the U.S. Air Force and Walt Williams was eager to take over the aircraft so the NACA could start probing the intricacies of aerodynamic heating. By early fall the handover was imminent, but the air force was reluctant to give up the X-2 without securing a new record. And so both parties worked out a deal wherein the X-2 would remain with the air force long enough for Captain Mel Apt to familiarize himself with the rocket-powered aircraft and make an attempt to reach Mach 3.

Men weaving through the ghostly mist of liquid oxygen swirling around a bullet-like airplane nestled beneath a towering mother ship under predawn skies was fast becoming a familiar scene at Edwards Air Force Base. It was the scene unfolding on the morning of September 27, 1956, when Captain Mel Apt arrived. The runway adjacent to the Rogers dry lake bed was buzzing with activity as technicians readied and fueled a small white aircraft with wide swept-back wings and a long pointed nose for the morning's flight. The mother ship was a B-50 bomber, its cargo was Bell Aircraft's X-2 Starburster, and Apt's challenge that morning was to become the first man to fly at three times the speed of sound.

The day's flight was Apt's first in the X-2. The original flight plan had called for him to keep his speed below Mach 2.45 and focus on flying the perfect flight profile, but the speed limit had been lifted in light of the NACA's pending takeover. And while he'd never flown the X-2, Apt had spent the better part of seven months preparing for the flight. He had studied performance and time data from previous X-2 flights and spent hours in the simulator. He had received multiple briefings on high-speed stability from NACA experts. He had practiced unpowered or "dead-stick" landings in an F-86 to simulate the X-2's gliding landing from altitude onto the dry lake bed and flown trial flight paths in an F-100 jet aircraft. He had performed ground runs of the X-2's engine to familiarize himself with its power and had worn his pressure suit for cockpit and failure procedures training in the aircraft. He was ready.

The Sun rose to reveal another bright and clear day in the Mojave. Shortly after daybreak the B-50 roared to life with its rocket-powered cargo snugly under its belly. The mother ship rose steadily, and at 31,800 feet the X-2 was released. Apt fell away from the mother ship and lit his rocket engine to quickly put a large distance between himself and the two F-100 chase planes monitoring the flight. He passed through Mach 1 at 40,000 feet and managed to fly a nearly perfect profile as he ascended to 72,000 feet where he nosed the aircraft over into a shallow dive. The engine burned slightly longer than anticipated. Apt reached Mach 3.2 before his engine cut out, its fuel used up. He successfully became the fastest man alive.

All that remained now was for Apt to make his gliding return to the Rogers dry lake bed. He knew from studying previous flights that the X-2 had to slow to below Mach 2.4 before he could safely turn the aircraft back toward the desert air force base; if he didn't, he risked the aircraft becoming unstable. But he also knew that his high-speed run had taken him quite a ways from Edwards. If he waited for the X-2 to lose speed before turning, he might not have enough energy to glide all the way back to the lake bed. Weighing his options, Apt ultimately decided to begin his return sooner rather than later. He began banking around and pitched his nose up at the same time so the full underside of the aircraft would act as a brake against the atmosphere. But he was still traveling too fast. Apt lost control and the X-2 started tumbling, battering its pilot against the sides and canopy of the cockpit. Subjected to 6 g's in all directions, Apt briefly lost consciousness before waking to find himself in a subsonic inverted spin at forty thousand feet, the X-2 spinning around its roll and yaw axes in opposite directions. He managed to separate the X-2's forebody from the fuselage but ran out of time and altitude before he was able to eject from the cockpit and ride his own parachute to safety. Apt was still in the cockpit when it slammed into the desert floor traveling at several hundred miles per hour.

Apt's flight was the last of the X-2 program, and the crash was a wakeup call for those pushing aviation into the hypersonic realm that highlighted a much larger issue. Though the problem began with the high speed turn back toward Edwards, the culprit had ultimately been inertial coupling, the sometimes lethal phenomenon wherein the inertia of an aircraft's fuselage overpowers the stabilizing forces of the aircraft, causing it to tumble. Once it starts to tumble, the pilot can't do anything to regain control until he reaches denser air. It was the same phenomenon that had nearly claimed Chuck Yeager's life in the Bell X-1A, and it was a problem that promised to only get worse when aircraft began flying at near-orbital altitudes.

The same day that inertial coupling claimed Mel Apt's life, the beginnings of a solution were already taking shape at the High Speed Flight Station. In a large, bright hangar alongside parked aircraft, tanks of compressed gas, and other miscellany was a large cross made of iron girders. Laying horizontally, it sat balancing with its center on a universal truck joint commonly used in automotive drive shafts. The joint acted as a pivot, allowing the cross to move freely in all directions. The strange-looking contraption was meant to replicate an aircraft in flight, the four corners of the cross corresponding to the four corners of an aircraft: the nose, tail, and the two wingtips. A month later, one of the High Speed Flight Station's newest additions, Neil Armstrong, sat in a seat on one of the girders' ends facing away from the pivot. In front of him was a board with three instruments displaying the cross's pitch and bank angles as well as its angle of sideslip showing the aircraft's angle relative to the oncoming airstream, effectively turning his seat into a rudimentary open cockpit. In his left hand he gripped a standard control stick that unconventionally combined his pitch, yaw, and roll control in one device. The stick was wired to six thrusters on the ends of the cross's limbs each powered by compressed nitrogen gas. With the flick of his wrist, Armstrong could send the gas shooting out from one of the thrusters, the force of which would move the whole cross on its pivot. Twisting the stick activated thrusters in the forward end of the cross to control sideways motions. Pivoting the stick forward, backward, and laterally activated the thrusters on the other three corners. At the end of each limb was a crash bar designed to reset the test if it hit the ground. Called the Iron Cross Attitude Simulator, for all its simplicity it was the first tool NACA pilots had to learn to fly at near-space altitudes.

For Armstrong, brow furrowed in concentration and eyes fixed on the instruments in front of him, the Iron Cross was his introduction to a whole new way of flying. He was just three years back from a tour in Korea with the U.S. Navy where he'd flown F9F Panthers, the service's first successful carrier-based jet fighter aircraft. This single engine straight wing fighter had a central yoke and pedals in the pilot's foot-well for rudder control. It was a standard cockpit layout, one that Armstrong had become familiar with over the course of a flying career that had begun when he'd earned a pilot's license at sixteen before getting his driver's license. Now he was learning to fly using only his left hand to control a stick that generated counterintuitive movements. Engineers rigged the Iron Cross such that pitching the nose up meant pushing the stick forward, an input opposite to a traditional stick control. The trickiest part, though, was figuring out how much thrust was needed to get the right response from the thrusters. And the displays weren't spectacular and hardly correlated to what he could expect in a real aircraft.

But it was exactly this type of flying that appealed to Armstrong. As early as elementary school he had already settled on aviation as his future career, though he dreamed of being an aircraft designer rather than a pilot. He viewed learning to fly as a means to an end. Understanding what it felt like to fly, he reasoned, could only help him design better aircraft. And so he set off on a self-guided quest to learn everything he could about airplanes, devouring both fiction and nonfiction about aviation. In grade school he built a wind tunnel at home, a project that added to his knowledge of aerodynamics and also taught him how to blow out fuses in his parents' house. Armstrong's formal pilot training began after high school when he won a university scholarship through the navy's Holloway Plan, a program that would pay his tuition fees, buy his books, and give him a stipend for room and board at an approved university in exchange for naval service. He enrolled at Purdue University, but toward the end of 1950 was forced to put his education on hold to serve in Korea. He returned in March 1952 with seventy-eight combat missions under his belt and spent five months ashore ferrying aircraft out of the naval air station at San Diego before leaving the navy to complete his degree.

Armstrong graduated from Purdue in 1955 with a bachelor's degree in aeronautical engineering, and he immediately sought employment with the NACA, drawn to the organization's blend of precision engineering and piloting expertise. His first choice was the High Speed Flight Station at Edwards Air Force Base, the research center where innovating aircraft were regularly flying innovative research missions. But unfortunately for Armstrong, his application ended up circulating among NACA centers before landing at the Lewis Flight Propulsion Laboratory in his native Ohio. Armstrong accepted the position and found himself working on anti-icing systems. Six months later, he learned that a spot had opened up at the High Speed Flight Station with Crossfield's departure. He packed up and drove out to the California desert.

When he first arrived, Armstrong was one of five pilots flying seventeen different aircraft. He began his tenure with familiarization flights, simple missions that would allow him to get the feel of a new aircraft before taking on the specific flight techniques he would follow on data-gathering flights. There were a number of experimental X-planes on Armstrong's roster as well as a handful of fighter aircraft designed to be flown through extremely precise flight profiles. After every mission, he would turn in the results of his flights to check that they lined up with the results of the veteran research pilots who had flown before him; it was the best way to measure whether or not he was getting the hang of it. He also flew larger planes such as the B-29 Superfortress that had air launched the X-1, a very different kind of aircraft. It was an especially exciting time for the young research pilot to arrive at the High Speed Flight Station, and as he gained experience Armstrong was gradually assigned more demanding missions in increasingly sophisticated aircraft. And training on the Iron Cross was preparation for his first truly exotic flight assignment.

Though broadly designed to teach pilots how to control an aircraft's attitude with one hand using reaction controls, the Iron Cross was specifically designed by NACA technicians to match the dimensions and inertia ratios of the X-1B. One of four second-generation X-1s built by Bell Aircraft, the X-1B was originally earmarked for research flights testing the aerodynamic loads on the airframe, but its purpose changed when it was handed over to the NACA. Three hundred thermocouples were installed on the X-1B, and beginning in August 1956 research pilots flew the aircraft on missions designed to gather data about atmospheric heating. But the NACA had plans to fit the aircraft with reaction controls. The Iron Cross was a way for pilots to get the feel for these controls in the safety of a hanger before taking the modified X-1B up into the upper atmosphere.

While Armstrong was learning to fly by reaction controls, the X-15's design was frozen. A full-scale wood and soft metal mockup painted black was hidden behind a walled-off area marked SECRET. For the men who had brought it to life, the sight of the mockup filled them with an overwhelming sense of pride. Everything fit, and the weight was within acceptable limits. To the untrained eye, the mockup was so precise it looked like it was ready to fly.

When Walt Williams, Scott Crossfield's former boss from the High Speed Flight Station, arrived at the Los Angeles factory to inspect the mockup on behalf of the NACA, he was as anxious to get his hands on the research aircraft as Crossfield was to take it for a test flight. Poking his head into the cockpit and touring around the fuselage, Williams fired questions at Crossfield, who knew the aircraft in such intimate detail that he could answer every one without hesitation. An air force officer also inspecting the mockup challenged Crossfield as to why there was no landing gear position indicator in the cockpit. Crossfield explained that when the X-15 was coming in for a landing at two hundred miles per hour with no engine power to make a second attempt, it wouldn't matter whether the landing gear was down or not; it was going to land. The wiring and cockpit indicator light simply wasn't worth the extra five pounds. The X-15 passed this first inspection with about a hundred requested changes, far fewer than the team had anticipated. The hypersonic aircraft could finally move into the construction stage.

By the end of the year, the next generation of flight research was on the cusp of taking to the skies with Armstrong training with space-age reaction controls and the X-15 coming to life. For the NACA, the air force, and North American Aviation alike, the aircraft couldn't start flying soon enough. Mel Apt's death not only highlighted the unknowns that continued to stand in the way of aircraft giving way to spacecraft, the loss of the X-2 marked the loss of the most advanced rocket aircraft. The X-15 was needed to fill this new void in research vehicles. But there was one question the X-15 couldn't answer and that was what would happen to men at near-space altitudes over a prolonged time frame because follow-up vehicles would take men higher into space for longer stretches of time, exposing the human body to radiation from space and a host of unknown problems. The X-15's flights would be too short to really probe the biomedical questions of what a human could tolerate on a high-altitude flight. But elsewhere in the country experts were devising ways to gather the human data the rocket plane couldn't.

CHAPTER TEN

The Floating Astronaut

Hardly anyone knew who John Paul Stapp was when he arrived at Muroc Air Base in the spring of 1947. The U.S. Air Force flight surgeon kept largely to himself, working closely with a skeleton crew of civilians he had brought with him from the Northrop Corporation out of Los Angeles. Stapp quickly gained a reputation for operating something of a black market on the base. With barely any funding or resources to bring their project to fruition, Stapp bartered, trading medical advice and examinations for the parts and equipment his team needed. However nontraditional Stapp's method, there was nothing nefarious about his actions. His project was sanctioned by the air force, albeit at such a cripplingly low level of bureaucracy that he was left painfully underfunded. The silver lining was that his low priority kept him off the radar of the higher levels, the men Stapp envisioned spending their lives safely behind large mahogany desks, men who he knew would penalize his failure but take credit for his success. Operating under their radar made him solely responsible for either outcome, but also afforded him great freedom.

Gradually and discreetly, Stapp's team built a sled reminiscent of a soapbox racer though this one was made of spare aluminum pieces bolted together rather than a wooden crate. It had a seat for a pilot and, appropriately for Muroc at the time but untraditionally for soapbox racers, a bank of rockets strapped to its back. The sled sat at one end of a two-thousand-foot-long track. At the other end was a braking system designed to stop the sled instantly, the sudden deceleration subjecting its unlucky occupant to a sudden high load of g-forces. This was the unorthodox contraption's purpose, to simulate what a pilot would experience in a high-speed crash. Textbooks said the limit of human tolerance was 18 g's, or eighteen times the force of gravity. Stapp's research was dedicated to determining whether, as he suspected, man could tolerate far more. The first step in his human deceleration research was this soapbox racer sled nicknamed the Gee-Whizz.

The Gee-Whizz first raced down its track with a 185-pound humanoid dummy in its seat. In December, after thirty-five test runs, Stapp took the dummy's place. One rocket was fired on the first manned test of his program, sending Stapp tearing down the two-thousand-foot-long track at a relatively easy ninety miles per hour. The next day, he was back in the Gee-Whizz when three rockets were fired, more than doubling his speed to two hundred miles per hour. Every time the sled came to a halt Stapp was exposed to a punishing load of negative g-forces when he slammed against the harness keeping him in place. The heavy bruising he sustained, along with abrasions, broken ribs, concussions, and even lost fillings, didn't deter Stapp. The tests left him battered but invigorated. Riding the Gee-Whizz, he survived a maximum load of 35 g's, thirty-five times the force of gravity, proving that men can withstand far higher forces of deceleration than anyone had previously thought.

News that Stapp was using himself as a test subject eventually made its way to his home laboratory, the Aeromedical Laboratory at Wright Field outside of Dayton, Ohio, though it was hardly a surprise. Stapp had a reputation for using himself as a test subject. He had once flown to forty-seven thousand feet in an unheated, unpressurized cabin just to experience the painful effects of the bends firsthand. It was a line of research that led to Stapp's discovery that a pilot could avoid the bends by breathing pure oxygen for a half hour before takeoff. But the rocket sled runs were too much for his horrified superiors at the Wright Field and, fearing he might kill himself, Stapp was ordered to cease sending humans flying down the track in his sled and use chimpanzees instead.

The NACA contingent at Edwards, however, was very interested in Stapp's human research. Being in the business of high-performance aircraft poised to push inexorably toward space, the NACA was interested in the human element of spaceflight, particularly the acceleration and sudden deceleration a pilot would experience when launching on a rocket. This interest prompted Stapp to defy his orders and reclaim his seat in the Gee-Whizz for the sake of the NACA's research.

By June 1951, volunteers had made seventy-four runs in the Gee-Whizz's nearly four years of operation, Stapp himself being its most frequent rider. But the sled at Edwards wasn't enough for the flight surgeon who wanted to reach faster speeds for more violent decelerations. Stapp's opportunity for a new phase of research came with his transfer to the Holloman Air Force Base in New Mexico, the site adjacent to the White Sands Proving Ground. Stapp arrived in New Mexico in 1953 and, still working with engineers from Northrop, built a far more powerful rocket sled called Sonic Wind No. 1. This sled featured a replica of a jet pilot's seat, a full propulsion section at its back end, and a simple but effective water brake system at the far end of the track. A scoop attached to the underside of the sled would dig into a series of dams between the track's rails. Resistance from the scoops digging into the water would stop the sled almost instantaneously for the hard deceleration Stapp wanted.

As was his way, Stapp used himself as the first human subject for the Sonic Wind's speed run. In March 1954, just six of the nine rear-mounted rockets ignited to propel the sled to the brief top speed of 421 miles an hour. It was a land speed record, and when the sled stopped Stapp was subjected to 22 g's. Unsatisfied, Stapp added more rockets and more aims to the program. Because the sled had no windshield, the faster runs were also a way to gather data on human tolerance to wind blasts, something else pilots would experience in high speed ejections. And he remained the test subject.

On December 10, almost seven years after his first rocket sled run in the Gee-Whizz, Stapp was strapped into the seat of Sonic Wind No. 1 as it sat at the end of its track at Holloman. His arms and legs were secured to stop them from flailing.

He wore a helmet, which was strapped to his headrest to protect him from whiplash. He had a bite guard in his mouth to protect his teeth. He also had a crew spread around the desert. Technicians running the test were on hand and photographers were ready to capture image data of the test, specifically the final deceleration. In the sky, air force pilot Joe Kittinger was waiting in a T-33 jet, prepared to fly over the end of the track at the moment of deceleration so the photographer seated behind him could capture the test from above. The timing for everyone was crucial, as were clear skies for the sake of images. Behind the sled's seat that morning was a cluster of nine solid fuel rockets that together produced forty thousand pounds of thrust. Stapp was only going to subject himself to this high-speed run once. Everything had to be perfect.

Once the morning clouds broke, the countdown for the test started. With a thunderous roar, the nine rockets came to life and in five seconds Sonic Wind reached its top speed of 632 miles per hour. Stapp was thrown against the back of his seat with the burst of speed, briefly losing consciousness. Overhead, Kittinger watched as the sled flew across the desert floor faster than his T-33 jet. Then the scoops dug into the water trenches, sapping the sled's energy and stopping it cold in just 1.4 seconds. The sudden deceleration subjected Stapp to 46.2 g's. He momentarily felt his body weigh sixty-eight hundred pounds as he slammed forward against his restraints with the same force as if he had smashed into a brick wall while driving his car at 120 miles per hour. And then the sled was still.

Immobilized by his restrains, Stapp felt unbearable pain. He was struggling to breathe. As emergency personnel removed him from the Sonic Wind's seat, he mumbled that he couldn't see, that he had somehow gone blind. Stapp was rushed to the hospital for a thorough medical examination, but amazingly the doctors found that he had sustained no critical injuries. He had cracked ribs, two broken wrists, burst blood vessels in his eyes, and minor damage to his circulatory and respiratory systems, but he was otherwise fine. After an hour in the hospital, Stapp had regained his eyesight and was eating a hearty lunch. Stapp wanted to push his rocket sleds further. Before he had fully recovered, he was already planning to add more rockets to Sonic Wind No. 1 with the goal of reaching one thousand miles per hour, fast enough to break the speed of sound. But the air force stepped in and grounded him from any more rocket sled runs for his own safety.

Though banned from further rocket sled tests, Stapp had other outlets for his curiosity. Wright Field was home to the Materiel Division of the U.S. Army Air Corps, the branch responsible for developing advanced aircraft, equipment, and accessories. This branch's activities made the Ohio site's name synonymous with developments in aeronautical engineering and innovative research. Not long after Stapp arrived in 1946, he had witnessed one such test on a warm day in June. Three men sat on the grass wrestling a crash-test dummy outfitted in a standard flight suit and cap into a simple seat. Once they had it strapped in tight, the dummy and seat were hoisted by a crane and lowered into the uncovered cockpit of an aircraft parked nearby. Six technicians ensured everything was in the right place before tying a rope to the back of the seat. The end of the rope was in the hand of another man who also sat crouching in the grass. The crowd of onlookers moved a safe distance from the aircraft, which in this case was just feet away. Then the man in the grass pulled the rope. In an instant, the seat carrying the dummy shot up and backward, tracing a high arc over the aircraft's tail. Both landed separately, the dummy in a net and the seat nearby on the ground. The landing was irrelevant; it was the explosive ejection that mattered, a novel concept pioneered by the Luftwaffe during the Second World War that was fast becoming a fundamental part of American aviation.

The test underscored something interesting for Stapp. The state of the art of aviation was advancing as rockets propelled aircraft higher and faster, but human pilots weren't getting any more robust. The disparity between increasingly powerful aircraft and human frailty became a research niche for Stapp. Deceleration tests were a part of the subsequent investigation, but tests of human tolerance to extreme environments was another pathway. Higher flights meant men would soon be visiting the upper reaches of the atmosphere, a poorly understood region particularly where human factors were concerned. These men would be dealing with radiation from space, exposure to a near vacuum, and undoubtedly psychological challenges.

Aircraft, Stapp knew, wouldn't be a suitable means to investigate the human factors of spaceflight. For all their technological complexity and sophistication, the rocket planes streaking through the skies over Edwards Air Force Base gave the pilot only brief exposure to upper atmospheric conditions before falling back toward the dry lake bed. Even the X-15's highest altitude flights would only expose the pilot to space radiation for a few minutes. The pilot would need a specialized flight suit, but he could essentially ignore issues of radiation. Balloons would put the pilot in the exact opposite situation. Rising to altitude slowly as the lifting gas in the cavernous envelope expanded, a balloon could theoretically stay at altitude for an extended period, exposing the pilot to space radiation long enough to gather data. Stapp earmarked balloons as the ideal test bed for this line of research. He imagined a balloon large enough to carry a pressurized high-altitude capsule beneath it that could serve as a floating laboratory. When Stapp arrived at Holloman in 1953, he established and assumed responsibility for the Aeromedical Field Laboratory, a major goal of which was to understand the hazards to humans in the upper atmosphere. Foremost were questions of cosmic rays coming from deep space and radiation exposure, and specifically cosmic ray bombardment of living tissue.

David Simons saw the same shortcomings in small, suborbital sounding rockets that Stapp saw in rocket aircraft flights. Not only did Simons want a longer exposure to the upper altitude, he needed a more reliable way to recover the biological specimens. The Alberts monkeys stood as a prime example. The first primates in space launched under Simons's guidance had fared horribly in the nose cones of their V-2

Blossom rockets. In seeking a means to study the effects of high-altitude flight on living beings, Simons also saw balloons as the best option.

Balloon flights had changed dramatically in the nearly two centuries since Joseph-Michel and Jacques-Ètienne Montgolfier launched a sheep, a duck, and a rooster in the open basket of a hot air balloon over southern France in 1783. And it was due in large part to Otto Winzen. After spending a large portion of the Second World War in an internment camp, the German-born aeronautical engineer was hired as the chief engineer at the Minnesota Tool and Manufacturing Corporation. It was here that Winzen was introduced to the world of ballooning. In late 1945, he was recruited by Swiss balloonist Jean Piccard to work on Project Helios, a joint program by the U.S. Navy, the National Science Foundation, and General Mills that would see a man launched up into the stratosphere. Winzen was ultimately recruited by General Mills to work on balloon development and establish the company's Aeronautical Laboratories, a branch whose usefulness persisted after Helios was canceled.

At General Mills, Winzen helped advance the science of ballooning. He developed a way of heat sealing plastic gores together and developed a load-bearing tape to seal these joints such that the weight of a balloon's payload was evenly distributed over its entire surface so the material was less likely to tear. He also developed a polyethylene balloon that soon took the place of rubberized ones. Though typically just one or two thousandths of an inch thick, the exceptionally strong plastic was resistant to expansion once the balloon was fully inflated, decreasing the likelihood of it bursting at altitude. Winzen's polyethylene balloons were also able to take advantage of a gas's expansion. A small amount of a lighter-than-air lifting gas could be pumped into the balloon on the ground. As it rose, the Sun's heat combined with the lower atmospheric pressure would cause the gas to expand to fill the balloon's full volume. Winzen launched the first balloon of his own design in September 1947. The next year he left General Mills to start his own balloon manufacturing company, Winzen Research, Inc., based in Minneapolis. The startup money came from his in-laws; his wife, Vera Habrecht, was the wealthy daughter of a society photographer from Detroit. It was Winzen Research that pioneered the use of polyethylene resin for plastic balloons that were opening doors for high altitude research at Holloman.

The first polyethylene balloon had been launched at Holloman on July 3, 1947, twenty days before the first missiles took to the skies over the military base. The years that followed saw balloons carry cosmic ray track plates to altitude to measure radiation. There were also flights using mice to determine the effects of cosmic radiation on living beings. The mice flights presented an interesting challenge for Simons, namely how to keep his animals alive in a pressurized capsule for longer flights than the Alberts' on V-2 Blossoms. Using the mouse as a yardstick, he developed a system wherein one "mouse unit" was the amount of heat produced by a mouse. From there, he could scale up his capsule designs for other payloads, for a menagerie of small animals rather than a large number of mice.

One day, Stapp walked into Simons's office wondering how many mouse units a man produced and whether it would be possible to launch a human in a scale-up version of his animal capsules. Simons thought over the question and did some basic calculations. The idea didn't immediately raise red flags for Simons. A five-hundred-mouse unit capsule should be able to carry a man into the upper atmosphere, he told Stapp, and keep him aloft long enough to gather data about the environment. Both men agreed that a human flight would not only be useful for biomedical research but necessary. Animal passengers couldn't do anything but breathe, and, if the mood struck them, also eat, urinate, and defecate during a high-altitude flight. Only a human could describe the experience, run tests, and serve as a subject for research into the psychological aspects of high-altitude flight and shed light on the thoughts and feelings future space travelers might have to leaving the Earth. It was an invaluable data set Stapp and Simons knew they couldn't get any other way. Besides, there was an increasing need not only to understand the upper atmospheric environment but to develop the capsules and self-contained pressurized environments that pilots would need in high altitude aircraft and space travelers would need outside the atmosphere. For the moment, a manned balloon flight was the best way to simulate the space environment in a controlled and sustained manner.

Heartened by Simons's likeminded thoughts, Stapp upped the ante. He didn't just want to launch a man, he wanted to keep him at altitude. He asked Simons whether it would be possible for one of these capsules to get a man to an altitude of one hundred thousand feet and keep him there for at least twenty-four hours. Simons said yes. Then Stapp asked if Simons would be willing to make the flight himself. For a scientist accustomed to being overshadowed by the fighter jocks who got the glory assignments, this was a rare chance for Simons to carry out his experiments in the actual test environment. He wouldn't have to rely on someone else's description of the environment, matching their retellings to the data to create a complete picture of the test. He would be the scientist and the subject all in one, reaping benefits from both sides. Yes, Simons told Stapp, he wanted to go up in the balloon. The new project underway, Stapp appointed Simons director of the Space Biology Program. Completing the trifecta that would send a man into the stratosphere, Stapp and Simons sought out Otto Winzen, hoping to bring his expert knowledge of balloons to bear on their program.

By the summer of 1955, the balloon engineer was partnered with the two air force doctors. The trio reasoned that a twenty-four-hour manned flight to 115,000 feet was feasible and would give them the data they were after. The air force also came on board, though with some reservations. The chief of the Human Factors Division at the Air Research and Command Headquarters provisionally signed off on the project in August of that year, providing the program didn't overlap with other balloon projects. One competing program was called Stratolab, a navy program funded by the Office of Naval Research and the National Science Foundation. It was an offshoot of the unrealized Helios program Winzen had briefly worked on to fly a laboratory above 96 percent of the atmosphere. The ultimate goal was to measure the near-space environment with almost no distortion from the atmosphere. The air force's own competing program was one investigating the biophysics of escape, specifically the physiological and psychological aspects of ejecting from an aircraft. This went hand in hand with Stapp's deceleration studies.

Stapp and Simons's project gradually solidified. They named it Project Daedalus. Daedalus, in Greek mythology, was a craftsman and artisan who created wings of feathers and wax for his son Icarus, whose hubris led him to fly too close to the Sun. But when it turned out that Daedalus was the name of a classified atomic-powered aircraft project, they renamed their ballooning project Man High, which was eventually streamlined to Manhigh. As planning moved forward, the program on the whole was hampered by meager funding and continued wavering interest from the air force. Though Stapp stood firmly behind the project at the local level at Holloman, he still needed to forward the proposal to air force headquarters for approval. Headquarters, unable to justify a manned balloon program simply under the heading of cosmic ray research, was not keen on the program. At the same time there wasn't enough reason to stop the program from moving forward either. And so Manhigh pressed on under Simons's rule as project officer operating on Stapp's all too familiar shoestring budget. Measuring the effects of cosmic rays on human passengers was the principle line of inquiry. Designing the capsule that could keep them alive through the flight was the program's second major research goal.

As Manhigh developed, the capsule proved to be the mission's biggest challenge. Scientists, Simons among them, had been launching balloons with animal passengers for years, but the human passenger demanded a far more complex system, as did the full-day length of the mission. The Manhigh capsule emerged as the most complex manned system designed to date, by necessity larger and far more precise than any animal system. The air force eventually signed a contract with Winzen Research in November 1955, and as the prime contractor, Winzen was responsible for building both the capsule and the balloon, and was also responsible for managing the development, fabrication, maintenance, and modification of the Manhigh capsule throughout the program's lifetime. And the time frame was short. The contract stipulated that the balloon be ready to fly at the end of January 1956. The first flight was tentatively set for March.

Project Manhigh's ambitious time line proved incompatible with its modest budget. Like Stapp had done at Edwards, he and Simons were forced to take whatever bits and pieces they could to bring to life the mission, and specifically the capsule. This piecemeal approach set them behind schedule. March came and went with no capsule and no balloon to show for the Manhigh program. Still, the program was taking steps forward. March brought formal approval from the air force command, though the endorsement demoted cosmic ray research to a secondary project goal behind development of a life support capsule system. As time wore on, the price tag on Manhigh rose steadily from the original projected cost of $29,950 to nearly $240,000 over the course of a year. The addition of launch support and flight tracking crews was partially responsible for the increase, but mechanical improvements to the capsule had raised the price far more. When cost overruns became more than the air force was willing to contribute, Winzen himself stepped in and assumed financial responsibility. Such was his and his company's dedication to seeing the project through.

Finally, in the spring of 1957, the Manhigh capsule emerged from its period of high-cost development. Made of an aluminum alloy, the hermetically sealed capsule was an eight-foot-three-inch-tall cylinder with rounded ends, held upright by tubular struts that doubled as a shock absorption system.

Close to the top was a band with six portholes. Each faced a different direction and some had angled adjacent mirrors so the occupant could have a full view of the environment around him.

The capsule itself was pressurized to an equivalent of twenty-six thousand feet with a chemically treated atmosphere that removed carbon dioxide and excess moisture. The capsule pioneered a multigas system: a combination of oxygen, helium, and nitrogen would provide a breathing atmosphere for the pilot and also decrease the risk of a fatal oxygen fire. The cabin was, after all, crammed with electrical systems that could short out and spark at any moment. And though life support was the program's principle goal, the possibility for disaster remained. As such, a personal oxygen system was included in the capsule should the cabin pressure fail or should Simons need to bail out from altitude, though jumping out and landing by personal parachute wasn't something he hoped he would have to do. Gathering the best scientific data meant a controlled flight from start to finish. If he had to bail out, he would lose the capsule and possibly the valuable data.

For constant communications, the capsule was fitted with a high frequency receiver so Simons could talk to the ground controllers. There was a telemetry system as a backup he could use to send messages in Morse code. This system used a series of batteries attached beneath the capsule, each fitted with a small parachute so they could be jettisoned as ballast. There was also a tape recorder on board so Simons could record his thoughts and impressions in real time, preserving his own visual, emotional, and psychological impressions as they happened. His vital signs, such as respiration and heart rate, would be monitored by onboard systems throughout the flight as well. The sheer number of electrical systems meant the Manhigh capsule didn't need a heating system. With all the instrumentation up and running, the capsule would be a hot, crowded space in which to spend a day. It actually needed a cooling system. The solution was an open container of water. Because water boils at a lower temperature at higher altitudes, an open container would draw heat from the capsule during the flight.

The completed Manhigh capsule was put through a series of unmanned tests to check out all its systems. Animal flights approximating the weight and oxygen usage of a man tested the capsule's atmospheric system. A crash-test dummy was dropped from altitude to test the personal parachute Simons would use in an emergency. Coming on the heels of these checkout tests, the first flight was shaping up to be a shake-down mission, one that would work out the remaining kinks before the real science missions could begin. Simons was ready, but Stapp had other ideas.

Stapp had done enough experimental tests during his tenure with the air force to know that first flights were notoriously plagued by technical issues. He didn't want Simons flying a checkout flight. He wanted this first flight to go to a test pilot, someone accustomed to testing new vehicles in strange environments who could react quickly in an emergency without losing his cool and sacrificing data collection. Someone, in short, who had a chance of surviving if everything went wrong. He solicited volunteers from the corps of test pilots at Holloman and found a very willing Joe Kittinger.

Stapp had first met Kittinger before he was flying a T-33 over the Sonic Wind No. 1's track in the New Mexico desert. Kittinger had volunteered for a mystery assignment that turned out to be one of Stapp's zero gravity projects; cursory research told the pilot that the flight surgeon was something of a mad genius who not so quietly believed humans could survive a trip into space. Once assigned to Stapp's project, Kittinger learned his task was to fly a jet in parabolic arcs, climbing at a steep angle then diving back down to give a medical officer in the rear seat brief spurts of weightlessness at the top of each arc. For Kittinger, the very precise flight profile was a fun challenge, something different that required a skill he didn't normally use when testing an aircraft. For Stapp, Kittinger's ability to fly a nearly perfect parabola on his first attempt was impressive.

The first time Kittinger flew a biomedical flight, Simons had been his passenger. The pair went up, and before long Kittinger became so engrossed in the challenging flight profile and Simons so enthralled with the novel sensation that they stayed up long enough to dip into their reserve fuel. Simons remained unfazed as Kittinger called down that he was going to make an emergency landing, completely confident in his pilot's skill as he watched firetrucks and emergency personnel racing to take their positions along the runway. Only when Kittinger lost an engine and found that his landing gear light refused to illuminate did Simons start to become concerned. He knew Kittinger was an ace pilot, but also knew that one engine and no landing gear was not a great situation. It turned out that the landing gear was down after all, and Kittinger managed a textbook landing with one engine. A stroke of good luck, he said.

Between theses parabolic flights and his participation in the Sonic Wind program, volunteering had paid off for Kittinger. So while he remained a pilot with the Fighter Test Section at Holloman, he found himself on loan to the Aero Medical Laboratory fairly frequently, his third time as the alternate pilot for Project Manhigh. Manhigh was something completely foreign to Kittinger. The capsule designed to simply hang below the balloon was a far cry from the precision flight paths he'd grown used to flying. But he had also noticed that on all the parabolic flights he'd flown he had never once been sick. It was something scientists at the Aero Medical Lab had noticed as well. They were convinced pilots would be the nation's first spacemen, and now Kittinger was a pioneer in that field.

Under Manhigh, Kittinger and Simons went through the same preflight preparations. Stapp outlined four basic tests each man would have to pass to qualify for a Manhigh flight. The first was a claustrophobia test. Not only did the Manhigh capsule barely afford its occupant any space to move around, the partial pressure suit severely restricted a pilot's movements. It was designed to constrict his body, and it was also deliberately too short in the torso to take into account the suit's expansion at altitude. This meant the pilot would be kept in an awkward, hunched over position during all prelaunch activities and the beginning of his ascent before the suit loosened up. Claustrophobia was the greatest potential psychological challenge, and so both Kittinger and Simons went through a twenty-four-hour test in a confined space.

The second test was a decompression test in a chamber simulating one hundred thousand feet, the capsule's planned peak altitude. Both pilots had to be physically and psychologically ready to react to a sudden loss of pressure. And because a loss of pressure might mean bailing out at altitude, both men also had to qualify as parachute jumpers. Kittinger did his training at the naval air station in El Centro, California, where a small group of air force and navy pilots tested new parachute designs. His first jump was so exhilarating that Kittinger stuck around and made nine more jumps that day, earning his navy parachute wings.

The final criterion Stapp had laid out was the most necessary one, that Simons and Kittinger both learn to pilot balloons and obtain their free balloon pilot's license. For the Civil Aeronautics Administration, this meant six instructional flights lasting two hours, one controlled flight to ten thousand feet, and one hour-long solo flight. Simons and Kittinger practiced in the two-man Sky Car training vehicle over farmlands in Minnesota. The duo learned how to manage ballast, dropping weight to lighten the capsule and ascend. They learned how to vent helium, losing buoyancy to gently lower toward the ground. It was a very different experience than jet flying for Kittinger and a very different learning curve, but the quiet stillness was peaceful and both men eventually learned the intricacies of balloon flight.

Throughout the training process, Kittinger asked thousands of questions and made almost as many suggestions. He was extremely invested in the program. He was, after all, putting his life in the hands of the Winzen engineers. He wanted to make sure they knew him as an individual, not just a name on a report. Simons, meanwhile, grew suspicious of Kittinger's motivation for getting involved with Manhigh in the first place. Still dedicated to Manhigh's scientific aims, Simons wondered if Kittinger, a newly avid parachutist, didn't just want to bail out and parachute down from Manhigh's peak altitude, even if the move meant compromising the success of the flight. His distrust of Kittinger growing, Simons finally confronted Stapp about Manhigh's inaugural flight, arguing that he ought to be the pilot. But Stapp remained firm in his conviction that the first flight go to Kittinger, the test pilot. That he would still be making the first research flight only partially mollified Simons.

Kittinger, for his part, approached his Manhigh flight more like a scientist than a fighter jock. Working with Stapp had instilled in him a fascination with the science and testing aspect of research flights, and as the flight neared he devoted an increasing amount of his time to the Manhigh capsule. He was determined to ensure everything stayed on schedule. With the launch just weeks away, Kittinger learned that the teamsters union at the Winzen plant was planning to go on strike. It was possible the plastic used to make the balloon might not be delivered on time due to the labor stoppage, ultimately delaying fabrication of the balloon and setting the program behind schedule. Unwilling to wait and see whether this worst-case scenario came to pass, Kittinger took matters into his own hands. Risking personal embarrassment and trouble for the air force, he hopped into an old C-47 military transport plane and flew up to Terre Haute, Indiana, where the plastic for the balloon was made. With a small cohort of conspirators, they loaded crates of balloon material into the plane and flew it covertly into Minneapolis under cover of night. From the airport, the crates were loaded onto a truck and delivered to the Winzen plant mere hours before the strike was scheduled to start.

With the material at the Winzen plant, the balloon came together under the watchful eye of Vera Winzen herself. She oversaw construction of a massive system of tables where the sixty long gores of plastic were heat-welded together before the seams were sealed with bands that would distribute the weight of the capsule evenly around the balloon. Her team of women then inspected every inch of the balloon looking for pinholes or tiny tears, anything that would compromise its structural integrity once inflated at altitude. The women worked in stocking feet and submitted to daily fingernail inspections to ensure they didn't tear the delicate plastic, a rigorous example of quality control Kittinger appreciated as he was trusting his life to the strength of this whisper-thin plastic.

About an hour before midnight on June 1, 1957, Kittinger had passed his final medical exam and was wriggling into his partial pressure suit at the Winzen research plant in Minneapolis. The capsule was also going through its final checks: its liquid oxygen system filled; chemicals added to the air regeneration unit; and the electrical and communications systems given a final check. An hour and a half later, Kittinger was sealed inside the Manhigh capsule, which took on the moniker of Manhigh 1 for this first flight. He left his faceplate open, taking advantage of the capsule's pressurized environment as nitrogen was bled from the cabin and replaced by helium, a gas that would reduce the possibility of Kittinger succumbing to the bends in the event of a sudden decompression, a potentially lethal sickness caused by nitrogen bubbling out of the blood. Kittinger half sat and half crouched in the capsule for hours as technicians ran a second series of checks before loading him onto a truck and driving the twenty miles to the launch site at Fleming Field near South Saint Paul. Still in the predawn hours, Kittinger watched through the small porthole windows as technicians buzzed around the capsule making the final preparations, backlit by the lights illuminating his capsule. As the Sun began creeping up above the horizon, an early morning fog swirled around the launch platform imbuing the scene with an eerie, surreal feeling. The whole team was there, Simons and Stapp included. After working for years with a shoestring budget, the men and women who had devoted their time, and in the Winzens' case, their own money, were finally seeing the results of their hard work.

As daylight seeped into the field, a gentle wind picked up, knocking the balloon around as it was filled with helium. Straining against the rollers holding it in place, the balloon was finally released at six twenty-three, carrying Kittinger off the launch platform. He wasn't alone in the sky; circling the area were photographic and tracking aircraft. But inside the capsule Kittinger was isolated. Exacerbating his solitude was a communications glitch with the very high-frequency radio. He could hear his ground crew but they couldn't hear him, forcing him to use the backup continuous wave transmission system. Tapping out messages in Morse code, an employee at the Winzen plant would receive the messages then call the ground crews in South Saint Paul to verbally relay the message. This roundabout method meant a delay in transferring critical information, but it wasn't a showstopper for Kittinger. He tapped out a simple message in Morse code: NO SWEAT. It was a trademark phrase of his that told ground crews he was fine.

The higher he rose the more the helium gas expanded and began to fill the balloon's two million cubic feet in earnest, drawing Kittinger inexorably further from the Earth. Less than an hour into the flight, Kittinger glanced down at his oxygen supply gauge and saw that he had used up almost half of his breathable gas in less than an hour, far more than was expected at that stage in the mission. Logic said to abort the mission then and there, but Kittinger's test pilot instincts took over. Not only did he want to complete the test for his own sake, he also wanted to justify the hard work of the entire team. Besides, he rather than Simons was in that capsule that morning to test the system, and test the system he would. Reasoning that he could probably make it to his peak altitude and back if he carefully rationed his oxygen usage, Kittinger began bleeding oxygen from his suit into the capsule's environment. Taking away his own personal backup survival system put his life even more firmly in the hands of the Winzen and air force personnel who had built the capsule.

Manhigh I rose steadily through the atmosphere, past the tropopause toward the stratosphere. At forty-five thousand feet, Kittinger experienced the full force of a jet stream. The

capsule was knocked nearly over on its side, and looking out
the portholes he saw the balloon was completely distorted
into a concave bulging mass of plastic. Putting the sight out of
his mind and his faith in Vera Winzen's army of women, he
waited for the balloon to carry him above the fierce winds. As
quickly as it had come, the force of the jet stream abated and
the balloon resumed its rounded form; Mrs. Winzen did
indeed make excellent balloons. Kittinger continued to
ascend, and as he passed seventy thousand feet, the world
outside his portholes began to change. Looking down he
could still see the familiar pale blue of the Earthly sky as a
thick band on the horizon, but as he moved his eyes upward
the blue quickly gave way to inky black. The Sun shone bright
and strong in the morning sky, but the sky around it remained
completely devoid of color. Kittinger was filled with a sense
of awe and privilege looking out his small windows.

On the ground, Simons became increasingly anxious as
the mission wore on. He was still convinced that Kittinger's
fondness for parachute jumping would win and that the pilot
would find some excuse to bail out, choosing a thrill over
science and sacrificing the capsule in the process. Impatient to
get Kittinger and the Manhigh capsule down safely, Simons
radioed the order to begin his descent just before nine o'clock
in the morning.

When he got the call, Kittinger knew he had enough
oxygen to stay up a little longer, and he wasn't keen to end
the mission just yet. Now at ninety-six thousand feet, he had
a view of the Earth that few people had ever seen, its curvature clear against a dark sky. And unlike the rocket plane
pilots flying high in the skies over Edwards Air Force Base,
he had the luxury of sitting and letting the view wash over
him in the silence afforded by his balloon's lack of propulsion
system. The sky took on a hue he had never seen from the
Earth. It occurred to him in that moment that he was the first
man to spend any length of time in the near-space environment, that he was, in a way, the first man in space.

Kittinger didn't have much time to savor the moment. He
was already venting the gas from his balloon to begin his slow

decent to Earth. But first he toyed with Simons just a little, playing off the doctor's obsession with Kittinger's fondness for parachute jumping. He thought a moment before tapping out his reply to the order to descend: COME UP AND GET ME. Simons was livid, convinced that Kittinger had succumbed to the "breakaway phenomenon," the theorized psychological condition wherein a pilot at altitude would risk his life trying to reach further into space. The thought was laughable to Kittinger, who knew pilots didn't actually suffer from this breakaway phenomenon, and Stapp, too, found Kittinger's reply humorous. Nevertheless, Stapp did follow Simons's lead in ordering Kittinger to begin his descent, to which the pilot tapped out the reply: VALVING GAS.

The sky outside the porthole windows gradually lightened and returned to an Earthly light blue as Manhigh 1 brought Kittinger slowly back down through the atmosphere. He carefully valved gas to shrink the balloon and decrease its lift and dropped batteries to lighten his weight and slow his fall. Opening portholes as he went to let atmospheric oxygen into the capsule, Kittinger landed neatly at twelve fifty-seven in the afternoon with his oxygen supply completely depleted, having traveled just eighty lateral miles from his launch point. Simons was the first to arrive at the Manhigh capsule and help a grinning Kittinger out onto the grass.

Considering it was a shakedown flight, Manhigh 1 was widely considered a successful mission. But for all the biomedical benefits of these controlled balloon flights, they were limited as well. Balloons could carry a man to a near-space environment but not all the way into orbit. For that, Kittinger would need to fly in a rocket plane or more likely on top of a rocket, and by the time he made this historic flight, the wheels were in motion for that latter route to space to take center stage. The time for slowly and methodically learning about space was fast coming to an end.

CHAPTER ELEVEN

Space Becomes an Option

For centuries, ancient and medieval Western scientists thought the Sun was a perfect, unchanging orb in the sky that moved around our planet with clocklike regularity. In 1543, Polish astronomer Nicolaus Copernicus tracked the movements of the planet Mars and found that the enduring heliocentric model was wrong, that the Sun was actually at the center of the solar system and the Earth and all the other planets circled it. But still the Sun's reputation as an unblemished body shining its light on the world remained intact. Then, around 1611, Italian astronomer Galileo Galilei used his telescopes to project images of the Sun onto a blank wall and observed dark spots on its face. These first blemishes on an ostensibly perfect celestial body forever changed our perception of the Sun's immutable nature.

Later generations of astronomers followed suit, aiming their telescopes sunward to find not only that sunspots are common, they also increase and decrease over the course of a regular solar cycle that peaks at a solar maximum every eleven years. By the mid-twentieth century, our once docile Sun was understood to be incredibly dynamic. In 1952, the impending solar maximum prompted the International Council of Scientific Unions to propose a coordinated research effort into various aspects of atmospheric science affected by the Sun's activities called the International Geophysical Year.

This wouldn't be the first time international scientists collaborated on a large-scale research project to understand the physical properties and processes affecting the Earth. More than seventy years earlier, Austrian explorer Carl Weyprecht proposed that the answers to fundamental meteorological and geophysical questions could be answered through a series of coordinated scientific expeditions to the

Earth's poles. Weyprecht died in 1881, but not before inspiring the First International Polar Year. Between 1881 and 1884, some seven hundred scientists from eleven nations established fourteen research stations in the Earth's polar regions and an additional thirteen auxiliary stations around the world. Their painstakingly gathered data were never truly utilized, but the model of a coordinated research program proved so sufficiently viable that it was resurrected in 1927 by the International Meteorological Committee. Another coordinated research program pulling together observations from Arctic and Antarctic stations was expected to answer lingering questions about terrestrial magnetism as well as auroral and meteorological phenomena, the types of things that now had immediate applications to marine and aerial navigation, wireless telegraphy, and weather forecasting. From 1932 to 1933, the fiftieth anniversary of the First International Polar Year, a second international cohort of men braved arctic conditions to establish research stations at the poles for the sake of gathering internationally useful data for the Second International Polar Year.

The outbreak of the Second World War disrupted this international scientific undertaking, and the results of the Second Polar Year lay untapped until a 1946 Liquidation Commission was formed to conclude all outstanding issues. And as international relations began the slow process of repairing after the war's end, international scientists became increasingly keen to renew contacts and resume joint research programs. A handful of collaboration-minded scientists proposed such a program to the Joint Commission on the Ionosphere in Brussels in September 1950, a third Polar Year from 1957 to 1958 to coincide with a period of peak solar activity. The proposal garnered enough initial interest to be passed on to a number of international research bodies, including the International Astronomical Union, the International Council of Scientific Unions, and the World Meteorological Organization. Each of these organizations was receptive to the idea but changed its focus. Consensus was that a geophysical research program would be far more useful in the long term than another polar program. This revised focus meant new research stations established extending away from the poles toward the equator. With this new emphasis, the International Council of Scientific Unions created the Committée Speciale pour l'Année Geophysique Internationale (the Special Committee for the International Geophysical Year) in late 1952, an international committee to oversee all aspects of the first IGY.

The CSAGI held its first comprehensive meeting in Brussels midway through 1953 by which time more than thirty nations had responded favorably to the idea of an IGY. From there, plans gradually, and somewhat painstakingly, started coming together. Basic guidelines included the overall directive that participating nations develop programs that would take advantage of the significant technological advances that had been made since the previous Polar Year. Each nation would also manage its own IGY program at a local level, though national planning committees had to take into consideration the various overarching scientific commit-tees' needs at every turn. But the nature of such a wide-scale program meant the details of the IGY were by necessity ironed out during days-long international conferences held in Brussels or Barcelona.

The United States' contribution to the IGY was starting to take shape in the spring of 1954. The prospective program was by and large in line with the overall IGY goals with one notable exception. In addition to ground-based research stations, American scientists wanted to explore the upper atmosphere using rockets. Initially, the American IGY program called for the use of rockoons, a rocket-balloon hybrid that lifts a payload to altitude by a balloon before launching it higher with a small rocket, as well as small Aerobee sounding rockets. These high-altitude payloads would carry instruments to measure atmospheric pressure, temperature, density, and also return data about the strength of magnetic fields and the phenomena of night and day airglow. Specific instruments would also measure ultraviolet light and X-rays in space, investigate the particles that cause the glowing aurora, estimate the planet's ozone distribution, the density of the ionosphere, and measure the effects of cosmic radiation. These rockets would launch on predetermined "World Days" of notable solar activity with the goal of gathering the most fruitful results. But scientists had been working with sounding rockets with mixed results for years. The larger scope of the IGY brought with it the possibility of using larger rockets.

For the scientists already working with upper atmospheric research, launching an Earth-orbiting satellite was not only a natural next step, it was also the best way to overcome the shortcomings of available technology. Small sounding rockets like the American repurposed V-2s launching from the White Sands Proving Ground could only gather data during the payload's few minutes at the top of its arcing trajectory. Sounding rockets could also only deliver their payloads to the upper atmosphere, not above the atmosphere, which meant that any data onboard instruments gathered about cosmic radiation or the space environment were not completely free from atmospheric disturbances. And because they are fairly small, no sounding rocket had the necessary power to launch any payload fast enough to send it into orbit around the planet. Balloon flights, just like the ones John Paul Stapp and David Simons were starting to work on at the Holloman Air Force Base in New Mexico, were similarly limited. Though balloons could keep a payload of instruments at altitude far longer than a ballistic rocket, they were also unable to escape the Earth's atmosphere.

Putting a scientific satellite in orbit would overcome these problems and explore wholly new regions of space all at once. A satellite could gather data on outer atmospheric density because it would be above the atmosphere and measure the Earth's equatorial radius and oblateness as it circled the globe. It was the only way to study the radiation environment outside the protection of the atmosphere. A satellite would travel around the planet at 17,500 miles per hour without ever falling back to Earth, extending its mission as long as its instruments had power to continue working.

From a practical standpoint, designing a payload to work in the upper atmosphere wasn't much more difficult than designing the same payload to work in orbit. In both instances the instruments would have to withstand the g-forces of launch, the cold of the upper atmosphere, and the vacuum of space. However, an orbital launch promised more demanding extremes for the payload and also required a power source that could keep the instruments working in these more extreme environments. The bigger challenge lay in developing the rocket that could propel the payload fast enough to get it into orbit around the planet, and that technology was no longer in the realm of futuristic fantasy. Military missiles under development could be modified to send modest payloads into orbit. As international plans solidified, more nations joined the effort leading to greater costs and loftier goals. Each participating nation was committed to using cutting-edge technology to investigate geophysical phenomena. And in many areas of investigation, such as aurora and other upper atmospheric and space phenomena, scientists unanimously agreed that an Earth-orbiting satellite would be the best way to gather the necessary data.

Support for the International Geophysical Year by necessity went beyond scientific circles. The American program was managed by the National Science Foundation at the behest of the chairman of the National Research Council, making it a government program that the president had to sign off on. The first time President Dwight Eisenhower was presented with the plans, he called the IGY a unique and striking example of international partners taking advantage of scientific curiosity in a way that promised to benefit nations worldwide. And the prospect of a satellite was equally appealing. The president first learned that putting a small satellite into orbit was scientifically beneficial and technologically possible during a meeting of international scientists in Rome in the fall of 1954 where he also learned that launching satellites under the umbrella of the IGY would make it a purely scientific endeavor. That the exploration of space be a peaceful undertaking was paramount for Eisenhower, something he knew wouldn't be seen as a competitive or hostile move to international partners. But he did realize that there would be an inescapable military connection to this program; the only vehicle then available that could get a payload into orbit was the army's Redstone family of missiles. Eisenhower approved the IGY satellite program on the condition that it not interfere with any ongoing missile program and that it not use a military missile—the launch vehicle would have to be some peaceful variant to firmly separate the military from space. The United States' provisional inclusion of a satellite to its IGY program prompted the International Council of Scientific Unions to urge other participating nations to consider building and launching small satellites as part of their IGY activities as well.

The Soviet Union answered the ICSU's call. Though not officially a participating nation in the IGY in 1954, it wasn't barred from taking part. The country had expressed interest in joining and was welcome to do so providing, like every other nation, that it freely exchange all gathered data with the other cooperating nations, something not forthcoming from a closed society. Regardless, Soviet representatives were present in Rome when the United States' satellite was approved.

Of all the scientists keen to pursue an orbital satellite for scientific research, perhaps none was more excited at the prospect than Wernher von Braun. After four years working at the army's Redstone Arsenal in Huntsville, Alabama, his Redstone rocket could be modified to place a small payload into orbit. Designed by von Braun and built by Chrysler, the nearly seventy-foot-long and six-foot-around missile was the U.S. Army's first short-range surface-to-surface missile. And with an engine capable of delivering seventy-eight thousand pounds of force at sea level, it was the best candidate to get a payload into orbit, though it couldn't do so alone.

Of all the V-2s launched in the United States, eight were designated Bumper rockets. These were ambitious variants of the German rocket similar to the unrealized two-stage A-9 and A-10 vehicles. Bumper was a V-2 with a Jet Propulsion Laboratory–built WAC Corporal rocket mounted on top as its second stage. The WAC Corporal's engine fired when the V-2 reached its maximum speed after the engine burned out. On one 1949 test, the combined momentum of the two rocket stages sent the WAC Corporal to an altitude of almost 250 miles. Considering the extra altitude gained from this second stage, it was clear to von Braun that multistage rockets were the best way to reach space. It was on this principle that von Braun based his first satellite proposal.

Von Braun's proposed program was simple. One of his Redstone rockets, too small to put a satellite into orbit on its own, would serve as the launch vehicle's first stage. The second, third, and fourth stages would be made of clustered Loki II-A rockets. These small rockets, measuring just a few inches in diameter, were also developed by the Jet Propulsion Laboratory for Army Ordnance based on the World War II-era German Taifun antiaircraft missiles. And the key to the design was in the staging. Like the V-2 Bumper, each rocket stage was designed to ignite when the one below it reached its top speed. The final stage was the orbit insertion stage, the one that would give a modest, five-pound satellite the final burst of speed it would need to achieve orbit around the Earth's equator at an altitude of about 186 miles. There were a few variants, one of which used nineteen Lokis in the second stage, seven in the third stage, and three for the final stage, but in any incarnation it would be a relatively inexpensive undertaking. Both rockets were available, so getting the satellite into orbit was simply a matter of putting the pieces together.

Von Braun took his proposal to the Office of Naval Research. The navy's pursuit of upper atmospheric research with the Stratolab had never come to fruition, and so von Braun suspected the service would be receptive to a joint program. Von Braun's courtship did spawn a joint undertaking by the U.S. Army and the U.S. Navy called Project Orbiter. The air force declined to join the program in part

because of interservice rivalries and in part because it preferred
to undertake its own satellite program rather than piggyback
on one designed by its former host service. Though no satellite program was formally sanctioned for the IGY, Project
Orbiter was formally proposed to the assistant secretary of
Defense in January 1955. And while it was a perfectly viable
satellite project, it was far from von Braun's dream program.
What he wanted was a long-term project that would create
something new, something purpose-built with longevity that
would also give his rocket team in Huntsville job security in
the longer term. By this point, von Braun had far loftier
missions not just in his mind, ideas he had already shared
with the American public.

Years earlier in October 1951, editors from the widely
circulated *Collier's* magazine were among the attendees of the
First Annual Symposium on Space Travel held at the American
Museum of Natural History's Hayden Planetarium in New
York City. One of the *Collier's* contingent was managing
editor Gordon Manning whose interest in the talks of spaceflight outweighed the skepticism of his reporters. Pursuing
this interest in space, Manning sent associate editor and
reporter Cornelius Ryan along with space artist Chesley
Bonestell to a space medicine conference in San Antonio,
Texas, months later. There, Ryan attended a talk by von
Braun during which the German engineer covered a blackboard with lengthy calculations and indecipherable schematics.
The room at large had gasped at these symbols and figures,
recognizing that von Braun had just demonstrated spaceflight's feasibility, but Ryan had no idea what the fuss was
about. He wasn't sufficiently well versed in mathematics to
understand the lecture without a translation.

Later, Ryan saw von Braun again as he was leaving another
session. Highball in hand, the editor grumbled that he had
been sent to find out what real rocket scientists thought about
spaceflight and all he was learning was that inexplicable
symbols on a blackboard sent conference rooms tittering in
excitement. Von Braun recognized the challenge facing
Ryan, that spaceflight needed to be broken down to the

layman. He also saw an opportunity to excite the American public about the realistic prospect of spaceflight. And so von Braun offered to bridge the gap between scientists and the public. A partnership between the engineer and the magazine was born.

Months later, the March 22, 1952, issue of *Collier's* appeared on newsstands around the country bearing the fruits of this chance meeting. The cover featured one of Bonestell's dramatic paintings depicting a winged glider atop a winged rocket at the moment of staging. Beneath the glider was the Earth's curving horizon giving way to the blackness of space. Above it, the feature article's tantalizing title read MAN WILL CONQUER SPACE SOON; TOP SCIENTISTS TELL HOW IN 15 STARTLING PAGES.

The article, written by von Braun, detailed the construction of an orbiting space station that would serve as an off-world science laboratory and jumping-off point for missions to the Moon. The vision von Braun brought *Collier's* readers became grander with each subsequent issue adding a new facet to space exploration. Over the course of two years, the magazine published eight issues unpacking his vision for the future of space exploration through a series of feature articles written by von Braun and other experts in the field, including Willy Ley, one of the founders of the Verein für Raumschiffahrt. In these pages, the average American learned about large multistage rockets, about satellites, and about the vehicles that would take men to other planets.

The articles caught the attention of Walt Disney, the cartoon magnate whose name was already synonymous with Mickey Mouse. Disney was in the midst of building Disneyland, an amusement park whose concept was so new that willing investors were hard to come by. But one deal Disney had managed to secure was with the American Broadcasting Corporation. ABC had given Disney money to build his park in exchange for an exclusive TV series of hour-long episodes showcasing everything Disneyland would offer future visitors. The series would follow the park's four theme areas of Fantasyland, Adventureland, Frontierland, and

Tomorrowland. The futuristic Tomorrowland area featured, among other things, a ride along a simulated Moonscape, which meant the related TV episodes had to deal in science fact rather than science fiction. Disney's production team reached out to Willy Ley, who in turn brought von Braun into the fold.

On March 9, 1955, some forty million Americans watched as Tinkerbell appeared on their TV screens, flitting about and waving her cartoon wand amid changing images as a deep-voiced narrator introduced the night's episode of *Disneyland*. A final, glittering wave of Tinkerbell's wand replaced an image of a castle similar to Sleeping Beauty's with a black screen. The episode title appeared in lowercase script, utilitarian in stark contrast to the glittering opening sequence: "man in space." Walt Disney himself introduced the episode as the camera panned over an office littered with rocket models and draftsmen standing at tables poring over papers ostensibly filled with calculations. Thirty-three minutes into the episode, after an introduction to rocketry featuring footage of Valier's rocket car runs and V-2 launches, America met Wernher von Braun. The engineer was introduced as one of the foremost proponents of spaceflight and head of the guided missile division at the army's Redstone Arsenal. His higher-pitched voice was marked by a slight but distinct German accent. His responsibility for the V-2 was mentioned but only insofar as it related to the development of America's rockets. The family friendly show made no mention of the V-2's sordid past and von Braun's reputation by association.

Over the course of three episodes, American audiences learned about spaceflight from von Braun. He detailed plans to build a toroidal-shaped space station two hundred feet in diameter circling one thousand miles above the Earth. He envisioned that this station would rotate to provide a low-gravity environment for its fifty occupants. From this station, von Braun said, astronauts would leave on missions to the Moon, ten-day orbital flights to explore it in unprecedented detail that would precede landing missions. But even more exciting was von Braun's vision for a mission to Mars, the subject of the third episode in the Tomorrowland series, "Mars and Beyond." And unlike his first envisioned flight to the Moon that would be a small scouting trip, this first mission to Mars followed the outline from his earlier *Das Marsprojekt* with a seventy-man scientific mission.

In the book, von Braun designed his Martian mission to start from Johnston Island, a tiny island 940 miles west and slightly south of Hawaii. This safely isolated spot would serve as a launch and landing point for orbital missions. With every launch, the first and second stages of these rockets would be left to fall into the Pacific Ocean where they would be recovered and towed back to Johnston Island by a tug. The third stage of the rocket would be a winged gilder, again reminiscent of the ultra plane. Only the winged stage would reach orbit, delivering its cargo and transferring any excess fuel into the Mars-bound vehicle before reentering the atmosphere and gliding to a runway landing on Johnston Island. The three stages would be refurbished and used for another launch, a conservation of hardware that meant just forty-six rockets and glider stacks could make the 950 missions necessary to complete assembly of the Mars spacecraft.

In von Braun's vision, ten ships would make the first foray to the red planet, seven carrying passengers and three only carrying extra supplies and fuel. Each manned vehicle would feature a sixty-five-foot habitation sphere providing quarters for ten men per ship and also carry 356.5 metric tons of extra propellant for the trip home. The three cargo ships, meanwhile, would each carry a two-hundred-metric-ton winged lander and an additional 195 metric tons of reserve supplies. From their starting point in Earth's orbit, the whole platoon would burn their engines for a full sixty-six minutes to gain the necessary boost of speed to reach Mars. The empty fuel tanks would be jettisoned to lighten the overall mass of the mission as the ships and their seventy intrepid explorers settled in for the 260-day transit to Mars.

Upon their arrival, the fleet would fire their spacecraft's engines again, this time against their direction of travel to slow down enough to be captured by Mars' gravity. Once in orbit, they would begin surveying the Martian terrain in search of an appropriate landing spot. Once a site was selected, a small advanced team of men would board one of the winged landing vehicles and separate from the rest of the group, deorbiting their spacecraft and gliding to a landing. Fitted with landing skids rather than wheels, it would likely have to land on one of the Martian poles to take advantage of the smooth surface layer of ice. Landing in a reliable region was vital; this first group's mission was a one-way mission in a spacecraft with no means for the crew to return to orbit. It would then fall to this small crew to establish landing facilities for the rest of the team still waiting in orbit. They would establish a small base camp and set up the runway.

The second landing party would bring the total population of Mars to fifty; the remaining twenty men would stay in orbit, monitoring the spacecraft and observing the planet. On Mars, living out of inflatable quarters and traveling around the area in the crawler vehicles, the landing expedition would spend four hundred days exploring the red planet, gathering samples, and running experiments. Once their sojourn was at an end, they would load their samples and anything else worth bringing back to Earth and board their return vehicles and launch from the surface of Mars into orbit. There they would reunite with their orbiting counterparts and board the main vehicles before abandoning the spent spacecraft at Mars. When they finally arrived back home, they would reenter the atmosphere in specialized gliders and land on a runway custom built for their arrival. From the first launch to the last landing, this expedition to Mars would last about 963 days.

While this Mars mission was a fantasy, von Braun knew that with enough money to work out the engineering problems and build both the launch sites and the flight hardware, his mission was not impossible. And though many remained skeptical of such a grand mission particularly in light of the fact that neither satellites nor men had yet gone into space, attendees at the space medicine conference had seen von Braun's rationale in action, as had readers of *Collier's* magazine and *Disneyland* viewers.

But von Braun's message of an exciting, spacefaring future with space stations and expeditions to Mars came with a warning. It was clear that any nation that figured out the intricacies of spaceflight and developed the powerful technology to realize this spacefaring future would be an unquestioned technological leader. Whoever conquered space first would have dominion in the sky and technological dominance over its adversaries. Von Braun urged the United States to devote the time and resources to putting a payload into space sooner rather than later, even if the payload was the five-pound satellite he was proposing as the payload for Project Orbiter. Because however feasible his space station and Mars missions were, these missions hinged on significant technological advances. A Project Orbiter–style program was only a first step in space, but a vitally important one. Expecting to put a small satellite in orbit by the end of 1956, the real success of this project for von Braun wouldn't be the technical feat but rather the psychological gain. The satellite would be a clear demonstration of the United States' prowess, the satellite serving as a reminder of that fact as it orbited the planet once every ninety minutes. It would be a small but important psychological coup. Project Orbiter was, in short, a fast way for the United States to assert dominance over the Soviet Union.

This early talk of satellite programs sparked some long-term thinking about satellites from the agencies in charge and begat a slew of committees who explored the potential behind these ideas. The Coordinating Committee on General Science reasoned that studies should continue with increased interservice cooperation before any one program was cleared to fly. The Ad Hoc Committee on the Scientific Satellite Program was tasked with preparing a paper for the National Security Council on a potential interservice satellite program and backup program with the CCGS as the organizing body.

The prospect of launching a satellite as part of the United States' International Geophysical Year activities formally moved from idea to policy on July 29, 1955, with an announcement from the White House. A press release issued the same day gave more details, clarifying that the project would be entirely scientific in nature, co-run by the National Science Foundation and the National Academy of Sciences. Supporting technical advice and help would come from the Department of Defense. Its experience with upper atmospheric research would be useful when establishing a launch facility, as would its access to military sites.

Though presented as a purely scientific undertaking, there was another, less overt facet to the American IGY satellite decision. Eight days before the White House's satellite announcement, Eisenhower had proposed the Open Skies Treaty at a summit conference in Geneva, Switzerland. Since the early 1950s, the United States had been trying to gain as much reconnaissance information as possible about the Soviet Union's development of offensive weapons, a significant challenge in light of the closed nature of the Soviet state. What surveillance flights the Americans did manage typically flew through international airspace off the coasts of Russia. The handful of flights that did cross into Soviet airspace was in violation of international law.

With satellites circling the globe in various orbits, it was inevitable that one would pass over some part of the Soviet Union at some point, and one misunderstood satellite could tip the fragile relations between the two nations and spark an all-out war. The Korean War had ended with a signed armistice in July 1953, and while it put an end to the fighting it failed to truly resolve the conflict, encouraging the continued American Cold War policies of Soviet containment and militarization. Keen to avoid a renewed international conflict, Open Skies was Eisenhower's attempt to level the playing field. He proposed that the United States and the Soviet Union freely exchange information on military establishments and allow aircraft overflights for the sake of verification. Such an exchange would ease fears of a surprise attack for both sides and facilitate the development of space exploration as a truly peaceful endeavor.

Open Skies, though highly regarded by European governments, was rejected outright by the Soviet Union. Soviet leaders feared it was an attempt by the United States to lull the nation into a false sense of security before launching a surprise attack. The rejection of Open Skies left the United States apprehensive of what long-range missiles and advanced nuclear weapons the Soviet Union might be hiding, a potential "Bomber Gap" that prompted Eisenhower to authorize continued reconnaissance flights through Soviet airspace, however risky. The IGY satellite program, however, presented a loophole. Primarily a scientific research program, satellites flying over the Soviet Union could either double as a reconnaissance system or mask a separate reconnaissance satellite program such as the air force's Project 1115 already under development.

By 1955, there were fifty-five nations slated to participate in the IGY, and interest in the American satellite program was rising internationally. During the Sixth Congress of the International Astronautical Federation in Copenhagen in August that year, the president of the IAF showed the *Disneyland* episode "Man in Space." Delegates in attendance were enthusiastic about the show and excited by the promises it made for the future. The Soviet delegates in attendance were equally interested in the program and asked to borrow the film to show it in their home country. It would be excellent to have the movie on hand for private demonstrations about spaceflight, they said, though they likely wanted it as much as proof of the emerging American technology that, with the help of German engineers, was close to solving the basic problems associated with manned spaceflight. Not to be outdone by the Americans, the Soviet Union announced at the meeting in Copenhagen its intention to also launch a satellite as part of the IGY. Reports began circulating of an ambitious eighteen-month time frame accompanied by promises that any Soviet program would better any American attempts, but few people took these promises seriously.

Now official, deciding between the various American satellite proposals fell to Donald Quarles, a member of the National Advisory Committee for Aeronautics, assistant secretary of Defense for research and development, and secretary of the air force. Quarles appointed an eight-man committee, two from each branch of the military service and two from his own office, that would weigh the nation's options and make a recommendation. The committee's chairman was Homer Joe Stewart from the Jet Propulsion Laboratory in Pasadena, California, giving the group the moniker of the Stewart Committee. Joseph Kaplan, chairman of the U.S. National Committee for the IGY was also on the committee.

The Stewart Committee recognized that Project Orbiter, backed by von Braun's stunning record of success with rockets, was the program most likely to succeed. Stewart himself had even been involved in the proposal. Coming from JPL, he had served as an analyst on the proposal and made two important suggestions that ultimately strengthened Orbiter's standing. He suggested replacing the Loki rockets with a larger solid-fueled rocket based on JPL's Sergeant missiles. Stewart had also suggested using the Naval Research Laboratory's miniaturized radio transmitter in Orbiter in lieu of larger army hardware, though this change never materialized. And the army-navy partnership also didn't last. Not long after Orbiter was first proposed, the navy withdrew from the project in favor of its own satellite proposal.

The navy's independently developed proposal was Project Vanguard, another two-stage orbital system. The Vanguard launch vehicle was a variant of the Viking sounding rocket developed in the late 1940s by the Naval Research Laboratory and built by the Glenn L. Martin Company, making it one of the earliest demonstrations of homemade American rocket strength. Developing a more powerful version of the Viking and adding upper stages would turn the small sounding rocket into one large enough to launch a small satellite into Earth's orbit. The navy pitched a second variant of this idea that would use a Viking first stage fitted with a Hermes engine and three upper stages making it powerful enough to put a satellite as heavy as fifty pounds into orbit as early as July 1957.

The air force also submitted two proposals that were, like Project Orbiter, based on its missile programs then under development. The first used the Convair-built intercontinental ballistic missile with an Aerobee-Hi sounding rocket second stage. The thrust of these two rockets combined could put a 150-pound satellite into orbit. The second proposal called for a Convair series B missile to launch a two-thousand-pound payload into orbit.

Although these navy and air force proposals existed on paper, von Braun's Project Orbiter was the only one circulating that was based on available technology, making it the lone system that could be built and launched within a reasonable time frame and for a reasonable cost. It was, as per his design, the best option if the United States wanted to get a satellite into orbit before the Soviet Union. Though the navy's Vanguard proposal used the Viking rocket as its core technology, developing the orbit-capable rocket would essentially call for building a new rocket. The air force proposals, too, were both based on Convair rockets that had yet to be built and proven reliable, making these programs almost too hypothetical to be seriously considered.

But technology and speed weren't the only considerations. The Stewart Committee also had to be mindful of other ongoing projects. Though Orbiter was the best option to launch a satellite in the immediate future, there was some concern that backing this program would divert resources and manpower needed to maintain the army's missile program. Likewise, a purely scientific undertaking from the air force also threatened to derail ongoing projects, notably Project Feedback under the Advanced Reconnaissance System. The service was already working on attitude, guidance, and control systems, a solar-electrical energy converter, intelligence processing methods, an auxiliary power plant, and was investigating the effects of nuclear radiation on electronic components for a satellite. Duplicating this work to separate it from the military was both inevitable and unnecessary. The navy's Vanguard was the lone proposal that wouldn't interfere with an ongoing program.

Another more pressing consideration was the effect a satellite launch would have on international relations. Even if it was a scientific mission, the show of technological strength would be inextricably linked. Here, Vanguard stood out— the U.S. Navy's proposal was the only one without any ties to Germany. For all of Orbiter's benefits, it was conceived of by an ex-Nazi engineer and both the Redstone and Loki rockets were based on his V-2 technology as well. The air force's missiles also used the V-2 as a jumping-off point. But the Viking missile at the core of the Vanguard rocket had been developed by the navy as a sounding rocket. It had always been destined for research, never for combat.

Vanguard alone promoted the idealistic notion that science was a peaceful undertaking. Backers in Washington appreciated the existing components, too, which promised to keep the program's overall cost down. Whether it would work was another matter, but for the moment, Vanguard's pedigree was best suited to the national aims of the IGY. It was the homegrown, peaceful, all-American exploration rocket with Viking's success rate backing it up.

And so the Stewart Committee put it to a vote on August 3. Members acknowledged that Vanguard was a complex system that would likely become more complicated and expensive as its development program wore on, but they also couldn't ignore Orbiter's ties to the Nazis. The decision was split with three votes for Vanguard, two for Orbiter, and two citing their unfamiliarity with missiles as cause to side with the majority. One committee member, Professor Francis Charles McMath from the University of Michigan Observatory, was out sick. His preference was for Orbiter, and had he been in attendance that day he might have swayed the two undecided members to land on the army's side of a split vote. But he wasn't, and the majority ruled in favor of Vanguard. The committee thus passed their recommendation to Quarles, adding the suggestion that the air force's Project 1115, Project Feedback, be retained as a backup. The Naval Research Laboratory would assume responsibility for the project with funding from the National Science Foundation. The Glenn L. Martin Company, which had built the navy's Viking rocket, would remain the prime contractor for the launch vehicle and its operation.

When von Braun heard the decision, he was shocked. He had been with his family in North Carolina when President Eisenhower announced the United States' commitment to launching a small satellite as part of its IGY activities and was certain that his own Project Orbiter would emerge victorious. After all, it was up against an unproven navy system and two air force proposals based on rockets that didn't yet exist. Finding that both programs were selected above his own was more than a frustration, and his ire only increased when he learned that his proposal might have won were it not for McMath's illness.

The news sent von Braun and Army Ordnance scrambling to revise the Orbiter proposal with the hope of changing the Stewart Committee's mind. The program was restructured to center around a larger rocket launching a heavier payload, effectively throwing out the window von Braun's original idea of a small satellite to score a psychological coup over the Soviet Union. A new, more powerful Rocketdyne engine added to the new Orbiter proposal promised a system that could deliver a 162-pound payload by late summer 1957. If the team succeeded in shaving seventeen hundred pounds off the Redstone, Orbiter would be able to get a more modest, eighteen-pound satellite into orbit by the end of 1956. And if reports of the Soviet Union's eighteen-month satellite program were to be believed, only the army's Orbiter could keep pace with and stand a chance at beating the adversary into orbit.

Von Braun did get a chance to try and convince the Stewart Committee to overturn its recommendation and choose Orbiter over Vanguard. He argued the army's readiness to launch and warned of the Soviet threat, urging the committee to consider what might happen if the Soviet Union were to launch a satellite first, and went so far as to promise a satellite in orbit within ninety days. But his efforts were to no avail. The majority still ruled in favor of the U.S. Navy's all-American Vanguard program.

The Vanguard program officially started on September 9, 1955, the preliminary plan calling for six vehicles for multiple launches. But before the year was out, von Braun's warnings looked poised to come true. The political and military situation was changing on an international scale as American intelligence uncovered the existence of larger Soviet missiles. American intercontinental ballistic missile programs became a national priority, as did gathering intelligence on Soviet missile activities using high altitude aircraft and spy satellites. The focus shifted for the Martin Company as well. Not long after beginning work on Vanguard, Martin won the bid to build the Titan I intercontinental ballistic missile. Now with a priority program on its docket, Martin moved its best people to the military project and left Vanguard all but orphaned. Watching on the sidelines as the national satellite program fell by the wayside, von Braun began taking measures to retain Project Orbiter, even if it wasn't a sanctioned move. So close to fulfilling his dream of reaching space, he wasn't going to give up without a fight.

CHAPTER TWELVE

The First Satellite Race

It was a warm day when Homer Joe Stewart arrived at the sprawling Redstone Arsenal in Huntsville, Alabama, at the end of September 1955. The visit was an unscheduled stop on his way home to Pasadena from Washington, a visit borne out of frustration. He remained convinced that the Naval Research Laboratory would fail to orbit a satellite on schedule for the International Geophysical Year, and he was determined to preserve the army's Project Orbiter in some capacity or other. Stewart wasn't alone in this conviction. Wernher von Braun shared the same concerns about the navy's Vanguard program, as did leaders of the Army Ordnance division. And so the two engineers devised a way to hide Project Orbiter under another name as part of another project, trusting the convoluted bureaucratic labeling system to act as camouflage.

Stewart and von Braun first revised the proposal, turning it into a more powerful launch system. Further analysis at the Jet Propulsion Laboratory supported Stewart's own recommendation to use scaled-down Sergeant rockets rather than the smaller Lokis as the upper stages. Eleven Sergeants for the second stage, three for the third stage, and one for the fourth stage mounted on top of a slightly elongated Redstone first stage promised a successful launch configuration. It was also one that could be modified to meet other program goals. An inactive fourth stage carrying a dummy warhead would make this rocket a perfect test bed for testing reentry materials. And so Orbiter was redesignated and retained under the guise of a reentry test vehicle, a small research and development program that would investigate how different materials and warhead configurations fared when reentering the Earth's atmosphere from a high altitude. Five missiles were diverted to this test program before the month was out, among them

Redstone missiles RS-27 and RS-29. If the navy failed and the need for a satellite launch became pressing, the program could easily be reconfigured to launch a payload into orbit. It wasn't an ideal solution, but Stewart was nevertheless satisfied when he left Huntsville.

Project Orbiter might have been temporarily saved, but the same could not be said for von Braun's team. Having lost the satellite bid, they were operating without any long-term programs, a situation that loomed as a threat to shut down the Redstone Arsenal altogether. Salvation came in November when, after the drawn-out deliberation over the satellite program, Defense Secretary Charles Wilson finally made a long-delayed decision on the nation's Intermediate Range Ballistic Missile programs. Wilson funded two missiles, an air force concept called Thor based on components of the Atlas missile, and a joint army-navy concept called Jupiter that would have to launch from both a ground site and a ship at sea. Securing a new program not only gave von Braun's team much coveted job security, it also brought an influx of personnel to Redstone.

The sudden increase in activity prompted the army to create a new organization separate from but on the same grounds as the Redstone Arsenal. On the rainy first after-noon in February 1956, the Army Ballistic Missile Agency was formally dedicated. It was a strategic move designed to cut down on the red tape surrounding missile programs by separating von Braun's group from the rest of Redstone. The new ABMA stripped resources and personnel from its former home. Under the direction of von Braun and the strong lead-ership of the chief of Ordnance, Industrial Division, General John Bruce Medaris, the group set about bringing more Redstones and the new Jupiter to life.

On the surface, Jupiter wasn't much more complicated than incorporating the latest developments in missile tech-nology into a new design. Fins, something von Braun had used in all his rockets for stability in flight, were becoming obsolete with the advent of increasingly sensitive control systems. Modern engines were not only more powerful, they were also able to swivel to help guide a missile in flight. The inclusion of these advances made von Braun's first design of the Jupiter missile an advanced intermediate range missile, but the navy disagreed. Because ships have a limited storage space, the navy needed the missile to fit inside a launch tube. And so von Braun shortened the Jupiter's body and added inches to its diameter to recover lost volume resulting in a fatter, squatter, finless vehicle with a large liquid-fueled engine in the rear. The missile's guidance system was another problem area for the navy. Ships constantly pitching and rolling in the ocean meant launching a missile at sea from an unstable platform while fuel sloshed around inside the rocket's body. Jupiter would need an extremely sophisticated guidance system to compensate for these launch conditions.

Pressure from the navy's unique demands was compounded by a directive from Washington in the spring ordering that some missile—any missile—launch on a demonstration flight as soon as possible. Reports of Soviet progress suggested the United States might be lagging behind on missile development. A demonstration flight would alleviate fears of a technological gap at home and abroad. But neither Jupiter nor Thor was close to launching or could be ready before the end of the year. Even the reconfigured Orbiter-turned-Redstone reentry test vehicle wasn't quite ready to fly. The rocket was in good shape, but engineers still had to figure out exactly what materials and design to use for the small-scale Jupiter warhead needed to justify the rocket's launch. But there was another option. Von Braun's team suggested that they launch the first of the reentry vehicles, Redstone RS-27, with a dummy fourth stage. Little more than a payload, the inactive stage would reach a range of several thousand miles to demonstrate the power of the missile system. It would be a placeholder launch unrelated to the Jupiter program, but it could launch as early as September. When, two weeks later, Medaris got the go-ahead from the secretaries of the Defense and the army, von Braun felt slightly vindicated. His unwavering commitment to Orbiter, even under a different name, and his crafty though sanctioned retention of the launch vehicle was paying off.

Though it was a suborbital flight, the basic Orbiter hardware finally had an official mission. And if it worked, there was a chance that the second vehicle, Redstone RS-29, could have a shot at launching a modest, seventeen-pound satellite into orbit before too long. Vanguard was, to no one's great surprise, months behind schedule, making it clear that some backup system would be needed before long.

As much as von Braun and Medaris wanted to use the capability they knew they had to launch an army satellite, they also had to be careful not to offend the navy team or appear to be positioning their system as a direct competitor to Vanguard. They also could not admit that they had circumvented national policy after losing the satellite bid. Interservice relations had to be maintained.

Redstone RS-27's potential as a satellite launch system was obvious, but it moved through launch preparations strictly as a nose cone test. The missile was redesignated yet again as a Jupiter C—Jupiter so the launch would be imbued with the support and high priority of that program, and the C denoting it as a composite reentry test. The Jupiter C, in reality, bore no similarity to the army-navy joint Jupiter missile. The only technological overlap was the payload, the 86.5-pound nose cone that would follow a high arcing trajectory to demonstrate that its aerodynamic shape and materials could protect it from the searing heat of reentry. The test was looking ahead to a time when the Jupiter missile would launch warheads so high that they would need to survive reentry to detonate on their targets intact.

In September, RS-27 was stacked and erected on its launchpad at the air force's Cape Canaveral launch facility in Florida amid a flurry of excitement from the army team eager to see their vehicle in action. But the energy in the air worried officials in Washington who suspected that von Braun might "accidentally" launch the test with a live fourth stage and put a small satellite into orbit. And so the Pentagon stepped in.

One day during launch preparations, von Braun got a call in his office from Medaris with an official order not to sneak a live fourth stage onto the rocket. Still wary, the Pentagon sent an official to Cape Canaveral to inspect the stage and confirm that a satellite launch was not von Braun's covert plan. These fears, though possibly justified, were unfounded. The RS-27's long Redstone first stage was topped by a live second and third stage but a dummy fourth stage. Rather than propellant, the upper stage was loaded with structural elements and sand as ballast. More crippling than the inactive stage was the missing attitude control system that would be needed to fire the upper stages horizontally on an orbital trajectory.

The Jupiter C stood on the launchpad as days of rain and bad weather pushed the launch later into the month. The skies finally cleared enough on September 19 to begin the countdown, and at one forty-seven the following morning the rocket lifted from its launchpad through the cool air. Watching the telemetry on their screens as it flew out of sight, ABMA engineers saw the Jupiter C performing perfectly. The short flight reached a top speed of thirteen thousand miles per hour and covered a distance of 3,350 miles downrange away from the launch site. And most importantly, the dummy fourth stage was lofted to a peak altitude of 682 miles. It was clear that had the fourth stage been active it could have sent a satellite into orbit.

Von Braun was elated by the flight that took him one step closer to space, but his joy was marred by the frustration of bureaucratic secrecy. He and his team were forbidden from discussing the RS-27 launch with anyone outside the circle of classified personnel. It wasn't too hard to see that this Jupiter C was a separate rocket from the joint army-navy Jupiter project that was facing a new set of challenges including a dissolving partnership. The navy was leaning toward switching to solid fuels, a safer and simpler option when storing a rocket on and launching it from a ship. This eventually became a deal breaker between the service branches, and the navy withdrew from the Jupiter program to pursue its own solid-fueled Polaris missile. Now alone on the Jupiter program, the army was left fighting with the air force over who would launch the Jupiter if it were ever actually built; assigning missile launch capacity to one service was a way to delete duplicate capabilities.

Secretary of Defense Charlie Wilson ultimately sided with the air force in a decision that severely crippled the army. The emphasis on missiles had prompted the army to slash other areas of its budget in favor of the Jupiter program. Suddenly, von Braun and the ABMA were left wondering if, after everything they had worked for over the last half decade, their program would be dead before it even left the ground in spite of the successful RS-27 launch. And all the while the Soviet Union was moving ahead with its own missile development program. Far faster than the United States intelligence had anticipated the Soviets had built the world's first intercontinental ballistic missile, the two stage SS-6 Sapwood rocket powerful enough to cover a distance of nearly five thousand miles. Its formidable range came from the four booster rockets strapped around the first stage, making it far more powerful than the most advanced missiles in the United States.

This ongoing interservice conflict over missiles ran in the background of Eisenhower's reelection campaign, though it was far from the most pressing problem facing the president. In the summer of 1956, tensions between Egypt, Britain, and France rose when Egyptian president Gamal Abdel Nasser announced the nationalization of the nearly century-old British-French Suez Canal Company; Nasser accused the outraged nations of attempting continued domination of a former colony. Britain took an aggressive stance, unopposed to using force against Egypt while also maintaining a military agreement with France and Israel to invade the Egyptian region. Israel acted on these plans, advancing to within ten miles of the Suez Canal in October.

Eisenhower stepped in lest hostilities break out between NATO nations and a Middle Eastern power that seemed poised to gain assistance from the Soviet Union and spark another war. The president's health was another issue. He suffered a heart attack in September, raising questions about his ability to see out a second term. But Eisenhower's popularity favored reelection. Throughout his first term his approval rating averaged 69 percent. The presidential election that fall ended up as a rematch between Adlai Stevenson and Eisenhower, and on November 6, Eisenhower again won by a landslide. The same day, the French and British governments agreed to a United Nations cease-fire agreement in the Middle East. It was a good day for the president.

While national and international events reached resolutions, the satellite question remained unclear and a fairly low-priority issue. The Martin Company had finished the Vanguard design in February, but the rocket was plagued by development problems. Von Braun remained hopeful that he would eventually get a chance to launch Orbiter. The air force was also confident that it might get a chance at launching a satellite if the navy failed.

The air force satellite reconnaissance program, Project Feedback, was left wanting for funding and support in the wake of the Stewart Committee's decision. To consolidate ongoing research, management of the satellite program was transferred from the Air Research and Development Command to the Western Development Division based in Los Angeles and was rolled into the development of an advanced reconnaissance system. This put all the air force's satellite research under Brigadier General Bernhard Schriever. Schriever supported satellite development on the condition that it not take away from the ongoing work on missiles designed to be used in combat. He ultimately knew that both technologies would be advantageous should the Soviet threat become more pressing, but funding wasn't forthcoming. Limping along as it was, the air force couldn't promise a satellite launch before 1959, which left it as an unlikely backup to Vanguard. Regardless, the service remained unwilling to abandon the project altogether. It was developing missiles and instrumentation that could turn a payload into a satellite.

Considering what an air force space program might look like seemed a worthwhile study.

Opinions were divided among various study groups considering ways the air force might take its first steps into space. Some focused on possible follow-up programs to the X-15. Though the closest thing to a flight-ready vehicle was the painted mockup at North American Aviation's facilities in Los Angeles, forward-thinking planners saw the potential of a hypersonic piloted spacecraft picking up where the X-15 was designed to leave off. The first step would be to send a man orbiting the planet in a capsule, something akin to a nuclear warhead that would ballistically reenter the Earth's atmosphere and be recovered by a parachute. This was a simple first step to see how a pilot would react to a weightless environment. A general purpose ballistic research program would come next, gathering data on how boost-glide vehicles might work as well as how to launch interplanetary and deep space probes.

The ultimate goal was to develop a manned orbital space-plane based on the antipodal bomber Walter Dornberger had pitched in 1952, the precursor technology to his ultra planes. Launched atop a rocket, this vehicle would carry a pilot around the planet before reentering the atmosphere where he would fly the glider to a runway landing. In November, personnel from the air force's Air Research and Development Command solicited the NACA to be a part of this follow-up program to the X-15. The NACA was amenable, initiating the preliminary feasibility studies to determine the hypothetical vehicle's design and time frame.

Other air force study groups focused on less lofty, more immediate goals for a space program that highlighted the military aspect. The air force's Science Advisory Board put together a space sciences study group that recommended a broader inquiry into questions of military defense in the space between the Earth and the Moon. Still other groups considered the potential of a purely scientific space program, one that would focus on gathering experience to ultimately lay a strong foundation for long-term planning in space. One early

1957 study run out of the Wright Air Development Center considered state-of-the-art technology ranging from structural and propulsive systems to electronics to human factors, with an eye on developing trends and possible future developments. What emerged from the exhaustive examination was a report, "An Estimate of Future Space Vehicle Evolution Based upon a Projected Technical Capability," detailing a sequential evolution of space vehicles.

The first step in this metered approach to spaceflight was the Expendable Earth Orbiter, Minimum phase. Small satellites weighing just one hundred pounds would be lofted into orbit using available rockets modified into multistage launch vehicles. Only the uppermost stage would demand significant technical advances, so this phase could launch relatively quickly. Simple orbital missions would serve as a test bed for the most basic technical developments like satellite operation, tracking, and precise methods of delivering a satellite into an exact orbit. These lessons would parlay directly into the second program phase, Expendable Earth Orbiter, Advanced. This phase would see not only significantly larger payloads weighing up to twenty-nine hundred pounds launched into space, but significantly more sophisticated satellites as well. These vehicles would be able to self-stabilize in orbit, collect data, and provide limited power for onboard instruments by a small nuclear reactor. Over the course of these two phases, instruments and satellite components were expected to develop to the point where miniaturization would allow scientists to pack more into these modest satellites, a necessary development for the program moving forward.

The Unmanned Recoverable Earth Orbiter phase would launch satellites into orbit using the same rockets as their predecessors, but this generation would use a drag balloon to decelerate and begin their fall back through the Earth's atmosphere, a parachute slowing its final descent. So singular was the focus on recovering a satellite from space during this phase that these satellites wouldn't send any telemetry back to Earth during the flight. Everything would be stored on board and read only after the satellite was recovered.

Having proved recovery from orbit was feasible, the next phase was Manned Recoverable Earth Orbiter, which would repeat the previous flights but with a man inside the spacecraft. The study expected that the pilot and all the necessary life support systems would increase the spacecraft's mass and size enough to necessitate development of a four-stage booster rocket, and this was still a fairly simple spacecraft. As with the first recoverable system, there were no parameters to include telemetry in these first manned missions. Instead, the pilot would give controllers on the ground updates via a voice communications system, while storing information on board the spacecraft that would be read after he landed. The byproduct of this phase would be the development of sophisticated life support systems, including a recycling atmospheric breathing system that could manage the spacecraft's pressure, temperature, and humidity. Other human comforts would be inspired by their military counterparts. Onboard meals, for example, would be similar to combat-type rations, a small variety of precooked concentrated offerings.

Gradually, these flights would teach the air force how to manage its manned spaceflight program, and with experience would come longer flights and more demanding missions. Eventually, these first space travelers and their managers on the ground would apply newfound skills to building an orbital space station resting some 1,237 miles above the planet with a total mass of about 669,000 pounds, more than half of which would be propellant to maintain its orbit. The idea was to build a station that would be constantly occupied by a rotating cast of crews. One crew would launch and take up residence in the station, performing various astronomical observations and experiments during their time on board while also learning how to make life in space more comfortable. When the mission reached its end, a new crew would launch armed with supplies and take the place of the departing crew who would return home in the now familiar manned recovery vehicle. Building the space station would necessitate major technological advances, particularly in booster technology, but would come naturally at this point in the program.

But the air force had no intention of limiting itself to Earth orbit for long. The Wright Center's proposal included seven vehicles designed to reach escape velocity, flying fast enough to leave Earth orbit, beginning with the Expendable Lunar Vehicle, Pass-By. Similar in mass to the advanced expendable Earth orbiting satellite, this spacecraft would use a far more powerful booster to achieve a faster launch speed, traveling directly from the Earth to the Moon. Pending advances in camera sensitivity, data handling, and accuracy of remote camera aiming, this satellite would carry a visual system to record the first ever close-up images of the Moon.

Following these initial flights, an advanced version of the satellite would be the first payload to attempt a landing on the Moon. This phase, again, would require significant improvements to the state of the art in terms of propulsion and particularly the propulsion needed to slow the payload for landing. In the summer of 1957, it was too early for the air force study group to detail what kind of science experiments and instruments should be on board this spacecraft, but doubtless it would include means of gathering measurements and analysis of the Moon's geophysical properties as well as its surface and atmosphere, if it turned out there was one. The next stage of the proposed spaceflight program, though only expressed as subtext at this point, was the goal of landing a man on the Moon in the same basic spacecraft.

However far into the future this proposal reached, in the summer of 1957 this incremental approach to spaceflight was neither too conservative nor too audacious. The first stages were well within the limits of existing technology under the assumption that advances would be made over the course of spacecraft and rocket development. But feasible as it was, the Department of Defense was wary of pursuing a military space program; missile development in the name of national security remained the most important goal. Air force activities in space remained underfunded and unsupported, the reconnaissance satellite program continuing to limp along in favor of missile development.

The air force's extensive studies into potential spaceflight programs underscored the assumption that space would be the dominion of this service. Being in the business of fast aircraft and powerful missiles, it seemed logical that eventually the air force would move from air into space. Industry partners seemed to agree, presenting other spaceflight proposals to the air force.

AVCO, one of the air force's largest contractors developing reentry vehicles for intercontinental ballistic missiles, submitted its first spaceflight study to the service toward the end of 1956. It focused on a pure drag reentry method. The spacecraft, a sphere slightly larger than three feet in diameter, would rely on a stainless-steel-cloth parachute whose diameter was controlled by compressed air bellows. This parachute alone would adjust the vehicle in orbit, begin its fall back toward the Earth, and slow its reentry.

Convair, the contractor behind the Atlas missile, presented its own spaceflight proposal to the air force in the summer of 1957 calling for miniature orbital test vehicles and reconnaissance satellites. The company envisioned a constant drag, blunted, conical vehicle that would begin its return to Earth from orbit by a small solid retrorocket system. A fiberglass-resin heat shield would protect the vehicle from the heat of reentry, and from an altitude of forty thousand feet the descent would be managed by a conventional cargo parachute. Convair also stressed the importance of learning about the human side of spaceflight, namely the challenges involved in safely recovering a man from orbit.

The Martin Company also presented its early research on satellites to the air force in the summer of 1957. Martin highlighted the military aspect of space exploration, culminating in a formal proposal outlining basic military requirements for a military outpost on the Moon.

In spite of detailed proposals for research and development programs coming from both military and industry studies, space remained something that appealed to only a small cohort within the service. On the whole, the air force remained committed to vehicles flying through the air, not those rocketing through space. The same was true for the army, whose space activities were relegated to continued development tests with missiles that could be used to launch satellites into orbit. But regardless of the expressed interest of both these service branches and preliminary programs, following President Eisenhower's decree the nation's first satellite program remained a firmly nonmilitary one resting on the shoulders of the U.S. Navy.

On July 1, 1957, White House Secretary James Hagerty announced the beginning of the eighteen-month International Geophysical Year. The most important result of the International Geophysical Year, he said in his address, was for participating nations to demonstrate their ability to work together harmoniously on projects that would benefit not one nation but humanity as a whole. He expressed his hope that this type of scientific cooperation would become common practice in other fields of human endeavor. With the IGY officially begun, it was open season for satellite launches. The military branches fighting for the first launch on the home front now faced the prospect of an international competitor beating them to it.

CHAPTER THIRTEEN

One Little Ball's Big Impact

Taking advantage of a moment of calm before Friday evening's activities began, Wernher von Braun ducked briefly back to his office at the Redstone Arsenal. He had spent the afternoon with General Bruce Medaris, head of the U.S. Army Ballistic Missile Agency, giving incoming secretary of defense Neil McElroy a tour of the Redstone facilities and the group's latest missile technology. The ABMA was, on the whole, thrilled to have McElroy replace Charles Wilson, a man whose popularity with the army had dwindled over the years. The Redstone group blamed Wilson's lack of foresight for their slow progress and fruitless pursuit of short-range missiles and were still seething over his decision to grant control of the nation's long range missile to the air force. But the ABMA was hopeful that McElroy might be more sympathetic to their cause and condone pulling the retained Orbiter capability out of storage, green-lighting their desire to serve as a backup for the navy's Vanguard program. In showing McElroy around Redstone all afternoon, von Braun had worked hard to sell the army, and specifically the ABMA, as the nation's best bet for launching an American satellite into orbit.

As Von Braun was preparing to leave his office that evening, October 4, 1957, for an on-base predinner cocktail party in honor of McElroy's visit, his phone rang. He picked up the receiver and heard a British voice on the other end ask what he thought about it. About what, von Braun shot back, clearly missing some vital piece of information. About the Russian satellite that the Soviet Union had just launched into orbit, the reporter on the other line answered. The news didn't entirely shock von Braun. As early as June, articles had appeared in the West saying that Soviet scientists had developed a rocket and all the necessary instrumentation to put a small Earth satellite into orbit. Intellectually, von Braun knew their missiles were large enough to double as satellite launch vehicles, but the reality was still somehow different. More than anything, he was disappointed that the decision makers in Washington hadn't seen the potential in these Soviet missiles that he had, that intelligence information and his own warnings of what a psychological coup it would be for the first nation to reach orbit has been ignored.

Arriving at the small cocktail party, von Braun sought out McElroy and Medaris, and news of the Soviet satellite spread through the crowd. The launch of Sputnik, everyone recognized, was a turning point in history. A stunned silence seeped through the room until von Braun broke it. Medaris watched as the frustration of Project Orbiter languishing in storage suddenly spilled out of von Braun in a vitriolic rant. He knew the Soviets had orbital technology, the German engineer said, and also knew that Vanguard would certainly fail. Von Braun turned his attention squarely to McElroy. The navy's program was too beset by problems to leave the ground. "Give me sixty days," he implored, sixty days and the army would get a satellite into orbit using hardware just sitting on shelves waiting to be put to good use. Medaris stepped in to temper von Braun's enthusiasm; the ABMA would need ninety days to launch a satellite, he said, but otherwise he didn't disagree with his rocket engineer colleague.

At the dinner table that night, McElroy sat between Medaris and von Braun, his meal marked by a constant stream of arguments in favor of the army's satellite program, punctuated with updates about the Russian satellite from a radio broadcasting throughout the base. Unwelcome as the news was to the engineers at Redstone that night, there was a silver lining. Having been beaten into space by the Soviet Union would doubtlessly break the chains keeping their orbit-capable Redstones on the ground. The next day, anticipating formal approval before long, Medaris told von Braun to take Redstone RS-29 out of storage and quietly start getting it ready for launch.

While von Braun and Medaris were dining with McElroy, President Dwight Eisenhower was at his farm in Gettysburg, Pennsylvania, when he heard that the Soviets had launched Sputnik (a diminutive term meaning "fellow traveler"). The 22.8-inch sphere weighed 184 pounds and was soaring overhead tracing one orbit around the planet every ninety minutes, broadcasting a short-wave radio signal.

The news of Sputnik spread rapidly through the country; radio and television reporters were as aghast as the public over the Soviet feat. It looked as though von Braun's predictions of the psychological impact of a satellite were spot-on. The nation was already in turmoil over the showdown between the Arkansas National Guard blocking nine African American students from desegregating a Little Rock high school followed by President Eisenhower's deployment of the Army's 101st Airborne Division to ensure the black students' safe entry into the school. Now the nation was rocked with a new fear of being at the mercy of the Soviet Union. Americans were forced to concede that the Russians could no longer be thought of as a technologically backward nation, an enemy against whom the wide Atlantic and Pacific Oceans provided a buffer.

Sputnik made the Soviets' technological prowess painfully evident. The upper orbital insertion stage trailing behind the satellite in orbit was visible to the naked eye, while two onboard radio beacons broadcast a steady beeping. An RCA receiving station picked up the signal that was then broadcast over NBC radio, recorded, and replayed for the nation to hear. The next morning's *New York Times* front page brought the news to anyone who might have missed it. SOVIET FIRES EARTH SATELLITE INTO SPACE; IT IS CIRCLING THE GLOBE AT 18,000 M.P.H.; SPHERE TRACKED IN 4 CROSSINGS OVER U.S. Throughout the nation other newspapers carried similar headlines saying that the Soviets had won the race. International newspapers went as far as calling Sputnik the Pearl Harbor of American Science.

The basketball-sized satellite was enough to tarnish the president's image. Detractors seized the event as proof of

Eisenhower's failure to heed warnings of Soviet missile strength. Of course, it had been Eisenhower's knowledge of both Soviet and American missile capabilities that had led him to make the original decision to favor missile development over space programs and raise the navy's effort above other satellite launching systems.

The scientific community had a decidedly different, nonpolitical reaction to the Soviet satellite. Most were elated, not just by the technological feat but also that Sputnik had effectively opened space as a new ocean for discovery. But at the same time, none could ignore the worrisome implication of Sputnik's weight. Small as it was, it dwarfed Vanguard's three-and-a-half-pound grapefruit-sized payload as well as the army's thirty-one-pound satellite, which meant the rocket that launched Sputnik was far more powerful than any American equivalent. A rocket that could loft something so heavy into orbit could carry a similarly sized or even larger payload like a warhead across the globe. It was this missile threat, not the satellite threat, that worried scientists. The difference between nations lay not in the technology itself but in the totality of the effort that brought that technology to life. It was clear the Soviets were putting their energies and resources into developing a new weapons system that could double as a satellite system, an approach that allowed for far faster development than the American decision to develop two independent and largely redundant systems.

On Sputnik's fourth day in orbit, Eisenhower met with his principle scientific and military advisers to discuss America's next steps. He was irked that reports of two U.S. Army officers griping about the Vanguard program had been leaked to the public, adding to the perception that he had knowingly put the country in a disadvantaged position. Because he knew the state of America's missile programs, the president was somewhat dismayed by the public's reaction of fear and panic over a small satellite that couldn't do anything but broadcast its beep—it certainly wasn't about to start dropping bombs from orbit.

The implications behind the rocket didn't inspire Eisenhower to change national policy regarding satellites, neither did the opinions of his gathered advisers. Among the assembled men was Deputy Secretary of Defense Don Quarles. Yes, the army could have launched a satellite the previous September had the fourth stage on the RS-27 missile been live, Quarles told the president, but he assuaged frustrations by rehashing the Stewart Committee's rationale behind favoring Vanguard over Orbiter, particularly the need to keep missile programs far more secretive than the International Geophysical Year programs, a secrecy that extended to include the launch vehicle. Quarles went on to point out a silver lining. Sputnik had just established the concept of freedom in space, and the small satellite might have cleared the way for Eisenhower's Open Skies policy.

James Killian, president of the Massachusetts Institute of Technology, was also in attendance that day, and like von Braun he was primarily concerned with how this simple satellite might warp the minds of Americans by instilling irrational fear in the public. He worried more than anything that Sputnik would heat up the Cold War, accelerate an arms race, or lead to the development of advanced military technology to effectively bring the world closer to the brink of another international conflict.

Having taken into account his advisers' opinions, Eisenhower ended the meeting with the order for McElroy to undertake a full review of the nation's missile program. The president also ordered outgoing Secretary Wilson to direct the army to ready a satellite launch system as a backup for Vanguard, just in case.

The next morning, representatives from the Vanguard program arrived at the White House to give Eisenhower a progress report. The program's director was there, John P. Hagen, along with William M. Holaday, the recently appointed director of guided missiles for the Department of Defense. The two men gave Eisenhower a fifteen-minute update on Vanguard. They stressed that test vehicle TV-2, which was being readied for a launch at Cape Canaveral, was an experimental rocket and not one that was equipped to place a satellite into orbit. Atop its active first stage were two dummy stages. The next test vehicle, TV-3, was also an experimental test vehicle, still in the factory being assembled. And though TV-3 was the first complete Vanguard rocket and would carry a small instrumented payload, it wasn't a satellite launch attempt. It was a mission intended to test the launch vehicle. If a satellite ended up in orbit it would be a great bonus.

This latest information in mind, Eisenhower held a press conference in the Executive Office Building before 245 gathered journalists. Greeting the room at ten twenty-nine, the president immediately opened the floor to the questions about Sputnik everyone had on the tips of their tongues. Calmly and deliberately the president addressed the media's queries, demonstrating extreme composure before a frightened nation. Yes, he admitted, the United States could have orbited a satellite over a year ago, but the cost of merging the satellite program with a missile program would have been a detriment to both scientific and military progress. No, he said, the air force's Strategic Air Command had not become a relic overnight. Yes, the nation's scientific satellite program remained a priority within the confines of the International Geophysical Year, still completely separate from national security. No, the Soviets couldn't use Sputnik as a launch platform to drop bombs on America. No, Eisenhower assured the nation, Sputnik did not raise his apprehensions about national security, not one iota. The American satellite program, he said, had moved forward since its formal approval in 1955 with additional funding going toward assuring a launch within the International Geophysical Year. It had never been the subject of a race with the Soviet Union. The navy had been working in the face of problems on the Vanguard program, and whether the other services knew of the navy's hardships or could have launched a satellite earlier he was unwilling to say.

The media inquiry continued. From the din of eager voices one asked about the Vanguard launch planned for December, whether it would be the initial proof of concept small sphere or whether it would be fitted with any instrumentation to gather scientific data. Calling back to his earlier conversation with Hagen and Holaday, the president answered vaguely that the original plan had been to launch a simple sphere; the instrumentation was expensive, so taking initial measurements of orbital velocity and direction was a cost-effective first step. There might be many of these test vehicles before any instrumentation launched into orbit.

Eisenhower didn't say that the next Vanguard launch would be an orbital attempt, but the press interpreted it that way, bolstered in this opinion by a press release accompanying the conference. As articles appeared bringing the latest from the president to the public, Americans learned that small test spheres would launch in December with fully instrumented satellites following early in 1958. There was no clear distinction between a test launch and a satellite launch attempt, a misinterpretation that effectively turned the Vanguard TV-3 launch into what Americans thought of as the first orbital attempt. This idea gained momentum with follow-up articles promising that America's answer to Sputnik would be circling the globe by the end of the year. The Vanguard team steeled itself for a push to bring their project to fruition, but one that did little to offset the ever-present uncertainties of the program.

The president's assurances soothed the jittery nation. It was the first step toward solving what Eisenhower viewed as the most immediate problem created by Sputnik, which was to give a perspective regarding the Soviet feat to offset panic. The second problem facing the president was a much larger and more complicated one. He now needed an accelerated satellite program that could match the Soviet accomplishment, but one that wouldn't appear to be a reactionary program. He was wary of reigniting the fear in the hearts of his nation anew.

Addressing this second problem became the topic of a meeting with the Science Advisory Committee of the Office of Defense Mobilization on October 15. Though scheduled before Sputnik's launch, the satellite served to refocus the day's agenda to address looming space-age developments. Eisenhower asked the committee, which included Killian, for their collective opinion on Sputnik. Specifically, the president needed to know whether these experts truly thought American science had been outdistanced. The consensus was that American science had not been completely outstripped, though the balance had certainly shifted to the point where an American response was necessary. The Russians had started picking up steam and seemed poised to easily surpass the United States, warned physicist Dr. Isidor Rabi of Columbia University. The best move, offered the president of the Polaroid company, Dr. E. H. Land, was to emulate the Soviet Union's apparent attitude toward science, which was to treat it as a source of knowledge and pleasure. Eisenhower was skeptical on this point, unconvinced that the Soviet system really valued the best minds and supported their rise to the intellectual upper echelons. But regardless of the situation in Russia, the president recognized the need to push science forward in his country, and to this point Dr. Rabi had a suggestion: anticipating an influx of policy matters centered around science, it might be wise to appoint a science adviser to the White House. The task facing the men gathered in that meeting was clear. There was no shortage of missile and satellite programs in the country from governmental and nongovernmental agencies alike. Now they had to figure out how to best support and even merge these existing projects in the name of progress. The next step would have to be vigorous and swift.

Killian sat in his office on the Massachusetts Institute of Technology's sweeping grounds a few days later when his phone rang. On the other end was Sherman Adams, Eisenhower's chief of staff. The president was considering creating a position for a special assistant to advise him directly on all things science and technology, Adams said, then asked whether Killian would be willing to travel to Washington to lay out the role of this new position. Though he didn't say it, the subtext behind Adams's request was an invitation for

Killian to fill this role himself. Days later, the president of MIT arrived in Washington armed with a memorandum outlining the benefits of moving the president's Science Advisory Committee from the Office of Defense Mobilization into the White House. This would create a direct line to the president and create a full-time member to the group who could deal with all matters of scientific importance. This advisory committee would have to be nonpartisan, Killian specified, offering entirely confidential and anonymous advice, effectively preventing any one man from leveraging this role into personal gain. It was Killian's opinion that associating science with his security and space divisions would protect Eisenhower from attack regarding his defense initiatives and policies. And the president's willingness to take council from science advisors, Killian thought, boded well for the future relationship between science and government in America in the post-Sputnik era.

The same day that Eisenhower met with his Science Advisory Committee to discuss the next steps in space, a long-planned NACA meeting began at the Ames Research Laboratory in California. Called the Round Three Conference, it was intended to discuss the next phase of high-speed flight research; it was the third round after first breaking the sound barrier with the X-1 and then exploring the supersonic and low hypersonic flight with the X-15. This third step would undoubtedly be higher and faster flights pushing into space. The fundamental question now became what that vehicle would look like.

Opinions were divided on the aerodynamics of this future hypersonic vehicle. Continuing in the vein of the X-15 with a flat-bottomed delta wing vehicle seemed like an obvious choice to some. This next step would be a piloted, hypersonic glider designed to launch on a rocket reaching speeds as high as Mach 17.5 at altitudes as high as seventy-five miles before gliding, unpowered, down to a runway landing at Edwards Air Force Base. Eventually, a faster launch would take this basic glider into orbit, giving it a dual purpose as a hypersonic research vehicle and the first manned spacecraft. This proposal was inspired in part by Dornberger's antipodal bomber, one that would build on the lessons learned from the X-15 and develop a new knowledge base.

One incarnation of this idea was Project HYWARDS, an acronym for hypersonic weapon and research and development system that surfaced in 1956. It was an air force project with supporting wind tunnel research from the NACA's Langley Laboratory intended to explore the potential of a flat-bottomed delta wing boost-glide vehicle capable of reaching speeds as high as Mach 18. It marked an enormous leap from where the X-15 would leave off, taking it inexorably toward space. The basic features of a HYWARDS-type vehicle were brought up for discussion again owing to Sputnik's launch at the Round Three Conference as one possible way forward in space. And because it had been proposed years earlier and gained preliminary traction among various study groups, many assumed the boost-glide vehicle would be the next step in spaceflight.

But support for a flat-bottomed glider wasn't universal. A number of engineers in attendance preferred a drastically different, blunt-bodied approach to the next phase of flight research. The idea arose as a by-product of missile research. Just as the X-15 returning from high altitudes needed some protection from the searing heat of reentry, so too did missile warheads. Typically sleek and reminiscent of rifle shells just like the rockets that launched them, these warheads had a habit of detonating harmlessly in the air above their targets. Aerodynamic heating from their fall through the atmosphere was enough to detonate the bomb. Seeing this problem, Harvey Allen, a high-speed research specialist from Ames, considered taking the complete opposite aerodynamic approach. A blunt body falling from the peak altitude after a launch, he predicted, would develop a bow-shaped shock wave beneath it, essentially creating a buffering cushion of air protecting it from the heat of reentry. Allen tested the novel concept with Alfred Eggers, a colleague at Ames, by placing miniature blunt missiles in a supersonic free flight tunnel, a sort of hybrid between a firing range and a wind tunnel. The idea proved sound, and while Allen and Eggers's report was released as a research memorandum, it didn't find a receptive audience in the early 1950s.

Allen and Eggers weren't alone in exploring the merits of a blunt-bodied vehicle. Former naval submariner Max Faget had arrived at the NACA's Langley Laboratory in 1946. That same year, the center created the Pilotless Aircraft Research Division in large part to satisfy the voracious research appetites of the rocket specialists, Faget among them. Over the course of his missile work at PARD, Faget developed the idea of a ballistic vehicle for spaceflight, one that would fall from orbit rather than glide to a controlled landing. He noted that any vehicle launched suborbitally, be it at hypersonic speeds or not, would follow a ballistic trajectory rising and falling at the same angle. This simplistic flight path negated the need for complicated avionics and control equipment because the payload was going to fall in an extremely predictive way. The same basic principle could be applied to a return from orbit, Faget said. Retrorockets could slow a simple capsule from orbit, making it fall following a ballistic trajectory, simplifying its overall return to Earth. The vehicle wouldn't have to glide or be controlled to a runway landing. Returning from space could be as simple as designing a capsule with enough inherent stability and structural integrity to survive reentry, while making it fall on a ballistic path. The spacecraft he envisioned was conical with a flattened bottom, eleven feet long and seven feet across at the base, weighing just two thousand pounds. It was simple and light enough, Faget said, to be launched on the larger missiles then under development in the United States.

Faget and Eggers compared notes at the Round Three Conference. For Eggers, the main problem was weight. A blunt body with enough inherent lift to follow a slightly more controllable semiballistic trajectory would be heavier than a simple ballistic spacecraft but would expose its occupant to less potentially devastating deceleration forces. Faget, meanwhile, held firm in defense of his ballistic design. But whatever details the men disagreed on, their main opposition came from proponents of boost-glide vehicles.

The question between blunt bodies and gliders aside, the overall feeling of those attending the Round Three Conference was one of mounting pressure to solve the atmospheric heating and reentry issue associated with missiles. Everyone there knew that the missile solution would eventually inform both manned and unmanned satellite reentry methods. It wasn't the immediate goal just yet, but the NACA was starting to think in terms of manned spaceflight.

Spurred on by Sputnik, some industry partners were also beginning to consider immediate paths to manned space-flight. The reality of the Soviet satellite hit Harrison Storms one night as he was leaving the North American Aviation plant in Los Angeles. Lighting a cigarette before getting into his car, he looked up to see Sputnik glinting overhead in the twilight sky. The X-15 was still in development and far from flying, but overseeing the project he knew its capabilities, and so did its chief engineer, Charlie Feltz. When Storms got home that night, he called Feltz with a radical solution to the new satellite problem: they could put the X-15 into orbit. It couldn't fly in space in its current configuration, but beefed up it could become America's first manned spacecraft. Storms conceded that it wouldn't be an easy job. Turning the X-15 into a spaceplane would demand new materials and a couple of big technological breakthroughs to manage the hypersonic reentry. But on the whole the vehicle was aerodynamically up to the challenge.

Inspired, Storms threw together a proposal for an orbit-capable variant of the X-15 called the X-15B, and, without stopping to undertake a detailed wind tunnel program, he took the proposal to Washington. His hastily assembled idea used two of North American's Navaho boosters clustered together to form the first stage with a single Navaho serving as the second stage. Atop this launch system would be the beefed-up X-15B, its own engine acting as the third, orbital insertion phase. The actual vehicle was larger than the first generation X-15, with room for two pilots. It was also heavier, boasting a thicker Inconel X skin for the hypersonic reentry. Otherwise, the X-15B was roughly the same as its predecessor.

Wind tunnel tests had proved the X-15's design was sound. Changing the basics for an orbital variant seemed like an unnecessary way to complicate the proposed program.

As he did with the overall design, Storms based the X-15B's flight profile off the early X-15 flight studies as well. The stacked rocket would launch from Cape Canaveral, the second stage taking over from the first after eighty seconds of flight. At an altitude of four hundred thousand feet, the second stage would fall away and the X-15B's own engine would ignite to accelerate the spacecraft and its pilots to an orbital velocity of eighteen thousand miles per hour. After circling the Earth three times, the pilots would use their engine to slow their orbital velocity and, following the same basic reentry profile as the X-15, glide from space to an unpowered landing on a runway at Edwards Air Force Base.

Excited though Storms was, the brass in Washington was not receptive to his idea. Storms's proposal might have been rooted in technology he knew would parlay into a viable spaceflight program, but it was almost too futuristic to be taken seriously. And his was also far from the only unsolicited spaceflight proposal to reach the nation's capital in Sputnik's wake. He returned to Los Angeles as one of 421 rejected spaceflight pioneers.

Though engineers and industry professionals entertained different ways into space, and the president worked to soothe his anxious citizens, the weeks that followed Sputnik's launch saw no signs of national fear ebbing. Unease over the American education system and a potential intellectual gap entered the conversation while security and defense issues remained popular topics.

The national panic over Sputnik got markedly worse when on Sunday, November 3, the New York Times front page brought more very unwelcome news: SOVIET FIRES NEW SATELLITE, CARRYING DOG; HALF-TON SPHERE IS REPORTED 900 MILES UP. The corresponding article was small, relaying what scant details were known to the American readership. Like its predecessor, Sputnik II, which was launched to commemorate the fortieth anniversary of the great October revolution when the Bolshevik party seized power in Russia, emitted a beep to be heard the world over as it traced an orbit around the planet. But that's where the similarities ended. This second satellite was six times heavier than Sputnik, the article said, and it carried a dog named Laika on board who, by all accounts coming from the Russian media, was still alive and well. The story continued in the days that followed with articles saying that Laika might be recovered and returned home alive.

The articles of Sputnik II's launch made reference to continued calm at the White House, but away from the public eye this second satellite was met with less calm than its predecessor. The satellite's weight of more than twelve hundred pounds spoke more to the Soviet's rocket power than the first Sputnik had, and the canine passenger was another source of concern. No American missile under development could hope to put something that heavy into orbit, let alone something as heavy as a life support system. Sputnik II strengthened the impression that the Soviets possessed a fully capable launch system and hinted ominously that a manned mission couldn't be far behind. The pressure for Vanguard's success increased while Eisenhower came under attack from Congress. Opposing Democrats in particular pointed to an ill-spent defense budget as the reason the Soviets had pulled ahead and left a missile gap in the wake of their orbiting satellites. It seemed as though the Soviets were fast outstripping the United States in military might.

Eisenhower addressed the latest space development on the evening of November 7, a day that had begun with newspapers carrying Soviet premier Nikita Khrushchev's boastful promises of a Soviet victory over the United States in heavy industry. At eight o'clock, Eisenhower sat behind his desk in the Oval Office and began an address on a live television and radio broadcast. "My subject tonight is science and national security," the president began, a topic he said was too large to be dealt with in just one address. After offering his congratulations to the Soviet satellite teams, Eisenhower began with the ongoing American satellite and missile programs. The first item on the agenda was the future of a scientific satellite as part of the International Geophysical Year. As a defense of the Vanguard program, Eisenhower reiterated the decision to separate the American satellite effort from the military missile program as a way to solidify the former as a purely scientific endeavor in line with the International Geophysical Year's goals. Months earlier, in May, the president told his nation, the decision had been made to test the satellite launch system in 1957 with the view of launching a fully instrumented vehicle in March of 1958. He stood by this decision and this time line, saying that the satellite program had never been a race and rather had always been a part of American scientific activities. The intent of an American satellite had never been to match or best the Soviet Union.

The president then laid out the next steps in the wake of the latest satellite launch. Foremost, he said, was to bring together the best minds the science community had to offer to address the space question. He announced the creation of the office of Special Assistant to the President for Science and Technology, a position Dr. Killian of MIT had agreed to accept and one he would fill with assistance from a strong advisory group of outstanding experts. Looking ahead at a larger-scale space effort, Eisenhower mentioned his agreement with the secretary of defense that any new missile or space-related programs would eventually fall under a single manager and administered regardless of separate services. In the meantime, the military would lead the way into space.

Vanguard was now at the forefront of national consciousness, an underfunded and highly experimental program bearing a nation's hopes for leveling the technological playing field on its fragile shoulders. And the country expected Vanguard to work the first time. But Vanguard wasn't carrying the nation's space hopes alone. The day after Eisenhower's press conference and five days after Laika launched into orbit, the Pentagon gave an elated von Braun the green light to launch a small satellite on a Jupiter C rocket. The strong separation between science and missiles had started to erode.

By the beginning of December, the Office of Defense Mobilization's Science Advisory Committee had been reconstituted as Eisenhower's own science advisory committee with a membership of eighteen. The newly appointed Killian served as chairman over this group, which included in its ranks Hugh Dryden serving as a member out of office from his station with the NACA. On the whole, Killian considered this group to be composed of remarkably astute, uniquely skilled scientists and engineers whose experience made them credible political advisers, exactly the men Eisenhower needed as close consultants. The group also shared a deep sense of responsibility to science and a firm faith in its importance to both the nation and its citizens. Killian felt the group was more charismatic than bureaucratic.

As the White House began incorporating science advising into its structure, the military's stance on space changed as well, most notably within the air force. In the wake of the Sputnik satellites, a sudden, feverish rush hit the service, and the air force's Air Research and Development Command sought to compile the pieces of a coherent space program that could recoup ground lost to the Soviets and also restore the nation's damaged prestige. The feeling of national urgency prompted planners to discard the slow, measured approach to spaceflight in favor of a crash program that could deliver immediate results. Committees formed, and one led by Dr. Edward Teller was charged with addressing the fundamental question of what the air force's immediate next step should be.

The men on the Teller Committee broke the question into four parts for their own internal dissection. They asked why the United States was in second place in the missile race, and if this was the case, what were the long-term military implications. Was it possible for a long-range program to restore national prestige? And regardless, what sort of short-term stunts could have an immediate but finite payoff that could restore the nation's image in the interim? The committee's conclusions were synthesized in the so-called Teller Report that offered few answers. It suggested that the years from 1959 to 1962 be treated as a period of national emergency with serious emphasis on using available hardware to complete space goals, effectively merging the satellite and missile programs. As for the space program itself, the Teller Report first recommended a program with short-term aims focusing on upper atmospheric research that could be done as part of the IGY. Following this initial phase, the committee said, the air force should pursue a long-term lunar program that would fly around or land a payload on the surface of the Moon, possibly leaving fluorescent powder visible from the Earth. Eventually, manned spacecraft would build on this initial foundation in space. The committee recognized that men would be flying inside those orbiting satellites before too long.

At the same time, Congress was also investigating the nation's preparedness to enter the space age, and the air force argued that space should fall under its jurisdiction as the natural extension from atmospheric flight. The air force also had the most experience working with and managing large-scale exotic programs like the X-1, a knowledge that would be vital to any space program without military exploitation. Exploration, the air force maintained, must precede exploitation. There would certainly be military benefits to exploring space, but none should be seriously considered before gaining a detailed knowledge of the new arena. This latter point was not lost on the Department of Defense, which was beginning to accept that research was not something to shy away from.

With the Sputniks soaring overhead, the air force devoted twelve research centers to studying various topics related to a future space program, including space medicine, propulsion, geophysics, communications, guidance, and test operations with the understanding that the Air Research and Development Command would eventually take control of the program. But the Air Force Ballistic Missile Command was also eager to grab the prize of space, though it preferred to skip over the short-term stunt phase of getting something into space as quickly as possible. Instead, it wanted to develop a lasting satellite program with real capabilities and a manned spaceflight system.

Against this background—the plethora of committees investigating America's space potential, changing attitudes in Washington, and the military branches—Vanguard TV-3 moved from the factory to the launchpad. The days leading up to launch were marked by surprisingly few problems. The second stage engine was replaced when a crack was found during an inspection. The first two stages passed static fire tests on the ground. The electronics and instrumentation were tested satisfactorily. All preflight operations proceeded without major hiccups. As November turned to December, thousands of people began migrating to Florida to witness America's first satellite launch. It was less than ideal for the Vanguard team who, in the early era of rocketry, would have much preferred to carry out a test of a new system without the nation and the world following along in person or by radio and television broadcasts. The national interest left the Vanguard team with no choice but to carry on and approach the launch as though the test had always been a satellite mission.

The rocket stood ready on the launchpad and the team was finally confident in their product, but the weather failed to cooperate. High winds and near freezing temperatures delayed the launch. The first attempted countdown was canceled on a cold Wednesday and started again on Thursday afternoon around five. The countdown proceeded slowly, pausing for a series of holds until ten thirty on Friday morning, December 6, when the countdown reached T-minus one hour; one hour to launch. The weather remained touch and go with gusts of high winds threatening to blow the small rocket off course at the moment of liftoff, but the Vanguard team pressed on. At T-minus forty-five minutes the men monitoring telemetry from the rocket got the all clear for launch. At T-minus thirty minutes, a blast from a bullhorn signaled it was time for crews to clear the launch area. The doors to the blockhouse where the launch team sat at consoles monitoring their rocket closed at T-minus twenty-five. The lights went out six minutes later. At the same time, a sign illuminated proclaiming the blockhouse to be a no smoking zone. At T-minus forty-five seconds, the electrical umbilicals powering from the rocket fell away, leaving Vanguard working on its own internal power supply. At T-minus one second, the test conductor signaled for a young Martin engineer to flip the toggle switch on his instrument panel.

Eyes all around the country were on Vanguard, the technicians in the blockhouse glued to the windows watching their test-turned-satellite launch. The gathered crowd at Cape Canaveral and those watching the live television broadcast saw sparks appear at the base of the rocket signaling the pyrotechnic igniter in place to light the engine was working. Oxygen and kerosene fumes spilled out of the engine and caught fire, turning the spark into a raging tempest of flames. The rocket shook as the thrust from the engine built up. Then, slowly, Vanguard started lumbering off the launchpad. It rose, but two seconds later started to fall back down. The men in the blockhouse instinctively ducked for cover. Outside, the rocket settled uneasily on its engines before falling against the firing structure. The tanks ruptured as the rocket toppled, spilling its fuel and oxidizer into the line of fire, sparking an inferno. In an instant, the pad was engulfed in flames. The fire-control technician instantly began the water deluge, emptying thousands of gallons of water onto the launchpad to quench the fire. It was a spectacular failure, and the world had seen it all. The crew in the blockhouse was dejectedly straightening back up when they noticed a beeping sound. The men were stunned as they realized the sound was the Vanguard satellite. It had somehow been shot away from the inferno and was triumphantly transmitting its signal as though it had reached orbit.

Vanguard's very public failure became, almost by default, a symbol of the American space program. And it garnered ridicule from around the world. International headlines in the days that followed offered new monikers for the small satellite including Puffnik, Flopnik, Kaputnik, and Stayputnik. The Soviet Union rubbed salt in the wound by offering assistance to the United States through a United Nations program that gave technological assistance to primitive nations. It was exactly the devastating blow to American prestige that von

Braun had predicted three years earlier in his initial proposal to launch a minimum satellite, and something he knew could have been avoided. Within the United States, Vanguard's failure had the effect of adding to the pressures already weighing on the president. Bowing in part to congressional and public demand, Eisenhower began to see that the country needed a centralized space program and an encompassing space policy; individual launches by the military would be fine in the interim, but it wasn't a long-term solution.

Killian, meanwhile, remained unwavering in his conviction that America would eventually succeed. He recommended to Eisenhower that Vanguard continue since one failure didn't speak to the program as a whole. Killian also recommended to the president that the Army be given full support and a launch window at the end of January. It was a silver lining to Vanguard's stunning public failure that thrilled von Braun. His team began preparing their system for launch. Formerly a Jupiter C, the Redstone first stage had been lengthened so the rocket could carry an additional propellant load. Eleven scaled-down Sergeant rockets were arranged in a circle making up the second stage, three for the third stage were nested within the second stage, and one making up the final orbital-insertion stage was stacked atop the third. These smaller rockets were positioned around bulkheads and rings, held together by an external shell making the whole stack sleek and smooth. Topping the stack was a small satellite named Explorer I weighing a little over thirty pounds, eighteen of which were the instrumentation and payload. Von Braun was ready, but he still had to wait for his launch window to open. For the moment, the navy's Vanguard was still the leading satellite program with another chance to get into orbit first.

Not to be left out, another air force spaceflight proposal surfaced in the wake of Vanguard's very public failure. The "Proposal for Future Air Force Ballistic Missile and Space Technology Development" put command of space activities squarely into the hands of the air force's Ballistic Missile Division that would work under the guidance of the Air Research and Development Command (ARDC). Supporting this future scheme, the ARDC Council in turn met with representatives from the air force Plans and Programming Office, the deputy commander for Research and Development, and the deputy commander for Weapons Systems midway through the month of December to develop a prospective five-year plan for the service's activities in space. Attending personnel agreed there was nothing more important than an all-out effort to develop America's space capabilities to overtake Soviet achievements, a program best defined as one that would address the major areas of reconnaissance, weapons delivery, space research and experiments, data transmission, countermeasures to potential Soviet aggression, and finally manned space flights. In reviewing existing technologies, a panel of experts agreed that the X-15 and a follow-up hypersonic boost-glide system akin to Project HYWARDS would be the best next steps in space because pilots would undoubtedly be involved. Human pilots would be vital players in the full exploitation of both the exploratory and military aspects of spaceflight, adept at reacting to new situations and able to manage new vehicles in strange environments better than any robotic system could.

In the wake of Vanguard's failure, it wasn't just internal proposals that were looking ahead at the air force leading the way in space. Half-formed industry proposals were reaching the service as well. AVCO and Convair pitched a joint program wherein an AVCO satellite would launch on a Convair booster. The Martin Company stressed the importance of establishing a manned base on the Moon as the air force's long-term goal and began floating such an idea to the service. The Aerophysics Development Corporation pitched a program consisting of a series of ballistic test vehicles developed alongside a lunar landing system. Bell Aircraft, the company behind the supersonic X-1, pitched a boost-glide vehicle. The Cook Research Laboratory wanted to construct and manage an air force–controlled orbital space station. General Electric wanted to work with the air force in developing a tracking station to monitor a series of orbiting satellites. Goodyear Aircraft wanted to build a shuttle system to ferry astronauts from the Earth to a manned orbiting laboratory. The Sperry Gyroscope Company wanted to undertake a research program wherein it would launch men on a trajectory that would put them into a state of free fall, gathering data about human factors of spaceflight in the meantime.

The differences between these proposals were minimal; each used a booster as the launch vehicle, and all but one put the satellite into an elliptical orbit rather than a circular one. All used retrorockets to return the vehicle to Earth with the exception of the AVCO concept, which would use a metallic parachute. But for all the contractor proposals, not one met the necessary requirements. Each industry proposal either failed to take into consideration the full range of human factors involved in spaceflight or failed to account for the whole variety of environments a man and his spacecraft were likely to encounter.

The silver lining for the ARDC, however, was that this influx of proposals helped narrow down what was important and what was superfluous when designing a spaceflight program. Weighing the proposals against one another did clarify favorable spacecraft configurations; a high drag capsule design with a flared-out bottom to hold the retrorockets, reaction jets, and recovery parachute was found to be the best first step. It quickly became clear that some kind of guidance system would also be vital to the first spacecraft to ensure orbital insertion and retrofire attitude. Without the right alignment the automatic retrofire burn risked sending the satellite away from Earth rather than beginning its fall back to the planet. Available intercontinental ballistic missiles would be the launch vehicles, modified as needed to meet the demands of a specific mission.

Some details were forcibly informed by existing knowledge. The landing methods were still up for debate, both runway landings and ocean splashdowns were attractive options. Materials, too, remained undefined, though the lightweight and strong Inconel X was a leading contender. The result of this investigation was the realization that manned spaceflight wasn't impossible given the state of technology. But what taking on a space program might do to the air force and its missile program was a separate question.

The NACA, meanwhile, ran its own studies of potential spaceflight programs at the same time as the air force under the guidance of Hugh Dryden. The NACA's participation served to widen the scope of the manned spaceflight project with the idea that the eventual program would be run the same way the two agencies comanaged the X-15. Each would bring its specific expertise to the problem at hand ultimately strengthening the product. But this arrangement also promised to make the NACA the governing body in spaceflight, something the air force wasn't keen on. For the NACA, the guidelines of its internal study were the same as those the air force followed, taking into consideration ongoing aviation and missile programs. Specifically, no space program the NACA backed could interfere with the X-15, the X-15B, or the hypersonic boost-glide program. These three programs were considered vital in their own right and important enough to continue to run in tandem.

This sudden burst of interest in manned spaceflight came to a head on January 7, 1958. More than 330 delegates from across the air force met at the Southern Hotel in Baltimore, Maryland, to discuss how the Air Research and Development Command might be reorganized to accommodate a new astronautics and space program. The meeting ended with plans that would see the air force's budget and long-term goals revised to include space goals through 1959, a commitment to programs that would produce viable results in this field, and a corresponding plan to limit the number of active programs to ensure success of those most important to the service's long-term goals. Ultimately, however, the new emphasis on space couldn't take away from the ongoing weapons development programs. At the end of the day, the air force was in the defense business and it would have to sell space to the Department of Defense if it wanted to see these programs get off the ground, and that meant not sacrificing the larger goals of the service.

At the same time, the air force's own multistage approach to spaceflight that had been passed over months earlier was suddenly back on the table. "An Estimate of Future Space Vehicle Evolution Based upon a Projected Technical Capability" was recast as the "Wright Air Development Center Long Range Research and Development Plans" and was modified to better address the Soviet Sputniks and incorporate a host of possible Americans' responses. The program reached the Air Research and Development Command headquarters before making its way to the Pentagon.

While the air force was talking about going into space, the army was taking its steps toward orbit. The Redstone RS-29, packed away more than a year earlier, was moved to Cape Canaveral and preparations for launch got underway alongside the second Vanguard attempt. Four times the Naval Research Laboratory counted down to launch another Vanguard satellite and four times weather or technical problems forced a launch cancellation while the army and von Braun sat by and watched. Finally, on January 29, the navy stepped aside and the army's three-day launch window opened. The rocket was erected on the launchpad, but its designer wasn't there. Under Medaris's orders, von Braun was in Washington, a personal frustration but a professional necessity. If the satellite did reach orbit, von Braun would need to speak at a press conference; his team could handle the technical end of things at the cape.

On the evening of January 31, 1958, von Braun arrived with a small cohort at the army's communications center at the Pentagon in Washington. The room gradually filled with military personnel, civilians, and representatives from the National Academy of Sciences' IGY committee. RS-29, renamed as Jupiter C, stood waiting 863 miles away, but the men in Washington couldn't see it. There was no closed-loop television system connecting the Pentagon to Cape Canaveral. Instead, a simple teletype machine projected updates on a screen and a small bank of telephones connected the room to the outside world.

The night wore on and the gathered men listened to radio reports as the launch was briefly held for a hydrogen peroxide leak. They watched as news appeared on the teletype screen. Von Braun anxiously read the progress reports: X minus one minute; X minus twenty seconds; X minus ten seconds. Then, finally: Firing commands; Mainstage; Liftoff; Program is starting; Still going. The lackadaisical updates were in direct contrast to the excitement building in the room. Ninety seconds after liftoff, the teletype related that the rocket had passed through the jet stream. The first stage cut off 156 seconds after launch and the teletype confirmed that the second stage had fired. But there was no confirmation that the third or fourth stages had come to life.

Von Braun had to wait for tracking stations to pick up the signal from the satellite for this vital information. He knew where Explorer I was supposed to be at what time after launch if it reached orbit, and the signals picked up by the tracking stations would confirm whether or not it had succeeded. The atmosphere in the army's communications center at the Pentagon was thick with tension as the men waited for an hour and a half, distracting themselves with coffee, cigarettes, and doughnuts.

The expected time to acquire the signal from Explorer I came and went with nothing but agonizing silence. Then, after eight tense minutes of quiet, nearly simultaneous calls came from four tracking stations on the West Coast confirming they had acquired the signal from Explorer I. The army had put a satellite into orbit. Instantly the mood shifted from worry to jubilation, bringing hardened military veterans to tears.

America had a satellite in orbit, but the diminutive Explorer I didn't quite level the playing field with the Soviet Union. The United States was still lagging in space, and exploring this new frontier with satellites and manned spacecraft promised to be a difficult challenge. Yet what was once the far-off future of starry-eyed engineers was suddenly around the corner, and everyone vying to build a satellite or run a space program knew that the next steps would set the tone for America's future in space.

CHAPTER FOURTEEN

The Fight to Control Space

Though he had shown remarkable composure when addressing the nation from his White House offices and had publicly denied any feelings of fear over the Soviet Sputnik satellites, President Dwight Eisenhower was privately relieved when he learned that the U.S. Army's Explorer I satellite was in orbit. Eisenhower had spent the evening at his cottage near the Augusta National Golf Club in Georgia, an evening punctuated by updates on the launch from his staff. It was his staff secretary general Andrew Goodpaster who told the president that the Jupiter C's second stage had fired. From that point on, like von Braun, Eisenhower had spent a tense hour waiting for confirmation on the satellite's fate. It was White House press secretary Jim Hagerty who had brought the welcome news on Explorer I.

The elated president tempered his excitement for the moment. Let's not make a big hullabaloo over this, he told Hagerty; Eisenhower was keen to avoid any backlash from a boastful pronouncement of American success. The small satellite was a step for the nation in space, but he also knew it was only the first in what promised to be a massive undertaking. The slow and methodical exploration of what lay beyond the atmosphere would be a costly venture whose success would hinge on a new way of managing such a large project. It was clear that sooner rather than later the nation would need a dedicated group to manage these activities.

One week later in early February 1958 Secretary of Defense Neil McElroy announced the creation of the Advanced Research Projects Agency (ARPA), a new agency within the Department of Defense that would manage all space initiatives with any intrinsic military value. As the head of this new agency, McElroy appointed an executive vice president from General Electric named Roy W. Johnson. Johnson in turn singled out the air force as the military branch most likely to manage a future space program, gaining and securing a strong foothold in this new field for the nation. The air force's Atlas missile then under development was the largest missile in America's arsenal and the best chance for launching a larger payload than the diminutive Explorer I satellite. But more than missile power, Johnson recognized that the air force was also the national leader in human factors relating to spaceflight.

Definitive proof of this had come on August 18, 1957, when Simons had climbed into the air force's Manhigh gondola at the Winzen Research plant in Minneapolis, Minnesota. By ten o'clock that night, he was moving swiftly through the prelaunch checklist. There had been some significant alterations to the capsule leading up to this second flight to correct the problems that had affected Joe Kittinger's mission. The oxygen regulation and communications systems had been largely overhauled, and a more sophisticated telemetry system had been retrofitted into the gondola to better monitor Simons's breathing, heart rate, and body temperature. The countdown moved forward. Thirty-seven minutes past midnight, Manhigh II was loaded on the back of a truck and Simons settled into the capsule for the 150-mile drive to the launch site in a deep open mining pit in Crosby. The flight surgeon managed only brief naps during the nearly five-hour-long drive; the dry ice cooling system on top of the capsule combined with the early morning chill made the gondola uncomfortably cold.

Launch preparations resumed once the capsule was at the launch site. The Sun rose over the pit, shedding light on the tense scene as minor problems threatened to cause major delays. As the balloon was slowly filled with a small amount of helium, reefing sleeves kept the plastic bubble contained and protected against a sudden gust of wind. But one sleeve tore loose, binding the balloon at one point some thirty feet above the ground. Knowing the material had to be freed before the mission could launch, Vera Winzen volunteered to climb a ladder solely supported by ropes and deftly cut the snagged band free with a pair of scissors. The situation resolved, the countdown continued. When ground winds got stronger, the crew picked up and carried the capsule and its semi-inflated balloon to the other side of the launch pit to ensure Simons wasn't blown directly into the nearby mountains.

Finally, after years of waiting and working behind the scenes on biomedical programs, Simons lifted off at nine twenty-two in the morning. He vented gas to control his ascent speed, all the while photographing and describing the view of the planet as it unfurled below him. Simons was mesmerized by the eerie quiet and gradually blackening skies, the likes of which he had never seen before. After two hours and eighteen minutes, Manhigh II reached 105,000 feet. From that altitude, Simons had had the sensation of bouncing like a ball in slow motion over an endless grid of farmlands crisscrossed by country roads, all following the curvature of the Earth.

Simons was just getting started on his astronomical observations when he got an unwelcome call from Otto Winzen. Manhigh II had lost its high frequency radio, Winzen said, meaning ground crews weren't getting any telemetry on Simons in the capsule. And, continued the German balloon engineer, it was likely that whatever had caused this failure would damage his voice communications system before long, too. Simons considered his options, whether or not he wanted to abort the flight while Winzen and John Paul Stapp waited for the doctor's decision. Simons reasoned that since he was physically alone floating more than one hundred thousand feet above the Earth, losing voice communications wouldn't have a marked effect on his isolation. Looking around the capsule, it seemed like all his other systems were working perfectly. Unwilling to throw away years of work and his one chance to pilot his own mission, Simons opted to stay aloft and gather as much data as possible. His scientific determination won out.

Though his systems were in good working order aside from the lost radio, Simons's flight soon took a turn. Planned to last a full day, the doctor was forced to extend his stay in the stratosphere when clouds rolled in over his landing area, threatening his descent. Simons conferred with tracking parties on the ground and in airborne craft, determining from multiple vantage points whether or not it was safe to attempt a descent. The clouds eventually broke and Simons was able to guide Manhigh II to the ground. He finally touched down just past five thirty in the evening in an alfalfa field near Frederick in northeast South Dakota. In spite of problems and an unforeseen elongated flight plan, the mission was deemed a success. And for his part, Simons was thrilled to have a working system on his hands that could keep a man alive for more than a full day in a space-like environment.

What Manhigh didn't have was a way into space; balloons could only float so high and couldn't give a capsule the speed it needed to go into orbit. Based in part on the strength of the Manhigh flights, the capsule approach to spaceflight was starting to gain favor over glider-type vehicles because of its simplistic technology. The air force's Atlas missile program was fast-tracked with the goal of consolidating ongoing developments that might eventually give way to a spaceflight program.

But the military emphasis in space tied up in the creation of ARPA didn't sit well with Eisenhower. He had thought long and hard over the best way to manage a space program and ultimately determined that peaceful exploration was paramount and the program should be directed by a civilian agency with no military ties. In the interest of international relations he knew that all purely scientific information about space should be shared openly and freely between nations, something that would be impossible with a military space program or even a science program with a strong military component. Keeping any large technological program a secret risked its being misinterpreted by the Soviet Union as a hidden weapons program, which could in turn see the Soviets develop an advanced secret weapons program of their own. It was imperative to Eisenhower that paranoia not develop into new weapons systems.

Although the president couldn't deny the military rationale of exploiting space in the name of national security with intelligence satellites and a military presence, his conviction that space be a peaceful, scientific arena weighed heaviest, and Eisenhower opted to keep a strong separation between the military and scientific exploration of his new region in the name of national morale and American international prestige.

On March 24, 1958, Eisenhower issued a memorandum to McElroy warning him that if a civilian space agency were to pass through Congress, it would take priority over ARPA. The president formalized this move with a request to Congress on April 2 that it establish a new agency to pull all existing space programs under one civilian umbrella. Recruiting Senate Majority Leader Lyndon Johnson to help pass the changes he thought necessary, the president's bill began the process of moving through the necessary channels to become law.

Defining what exactly this new space agency should do fell to James Killian, Eisenhower's science adviser. Just days after Explorer I reached orbit, the president had appointed Killian to a President's Science Advisory Board panel tasked with outlining the goals for and management of this new civilian space agency. Under Killian's guidance the panel considered the programs already in existence throughout the country, and in light of Eisenhower's partiality to a civilian group determined that the NACA was the best option. This long-standing aeronautical institution was unique in that it didn't answer to the president but instead reported to a board of directors led by a chairman. And though this board did include military representatives, they and their military goals were tempered by the cadre of scientists that served alongside them.

This cooperative relationship was echoed in the NACA's proven track record of working with and bridging the gap between military and civilian clients, an arrangement that typically allowed for shared research to benefit a single program like it was doing with the X-15. It was a way of working that promised to translate nicely to space. But without a new, forward-facing space goal added to its agenda, the NACA would be left duplicating military efforts in building rockets and aircraft. Killian's panel also saw the value in keeping the best minds in aviation under one organization umbrella as opposed to having them spread throughout the military and industry partners. Retaining the NACA would consolidate the nation's best minds, focusing them on the problems of spaceflight. The NACA was also a nonmilitary agency, satisfying Eisenhower's requirement that space be a peaceful undertaking.

As Killian saw it, the NACA could be easily adapted to incorporate a space program without sacrificing the ongoing research that was continually yielding advances in the still-developing field of aeronautics. From its humble roots with one small site in Langley, Virginia, the NACA had grown to include the Lewis Laboratory in Ohio and the Ames Laboratory and High Speed Flight Station at Edwards Air Force Base in California, all managed by Director Hugh Dryden in Washington. And each of these sites was dealing in some way with the technology that would eventually be applied to spaceflight, from powerful engines to materials able to withstand the searing heat of atmospheric reentery and exotic shapes for manned satellites.

The NACA's assets totaled some three hundred million dollars in research equipment and laboratories and nearly one third of its staff of eight thousand men and women were highly skilled engineers. Expanding these research sites to include space was a far simpler prospect than establishing all new sites for similar research. On this point the Bureau of the Budget added its approval. The NACA's average annual budget of about one hundred million dollars would have to be increased to bring spaceflight to life, but it would be a far cheaper alternative to establishing a wholly new organization with new research sites.

Expanding the staff and offices would be a challenge, but it would be a small step in the overall goal of developing a space agency. But the difficulties, agreed the members of the President's Science Advisory Board, could be fairly simply over-come by amending the laws governing the NACA to allow the organization the growth it needed to solve the problems.

A Government Organization Committee appointed by Killian weighed all options. It decided against handing the space program to the Atomic Energy Commission because of that body's singular focus on atomic energy in a field where chemical energy would surely be more useful in the short term. The committee finally favored the NACA over ARPA, citing the ultimately limited use of space by the military over scientific exploration.

The task of defining this new space agency also fell to Killian, which he delegated to an appointed panel of PSAC members chaired by Edward Purcell, a Harvard physicist and Nobel laureate. The Purcell Committee's conclusions came in the form of an essay called "Introduction to Outer Space," an essay that so eloquently detailed the rationale behind and technical challenges of spaceflight that Eisenhower read it to the nation during a broadcast from the White House on March 26. Nothing included in the statement was science fiction, the president said; it was a sober and realistic presentation of facts. The first steps would be straightforward scientific exploration, Eisenhower told the nation, studying the physical aspects of unmanned spaceflight going as far as the Moon. From there, the program would expand to send probes to distant planets with the expectation that humans would follow before long. And while there would certainly be a military aspect to the nation's future in space, the president assured the country that any such activity would not endanger the nation's security. National standing was the foremost consideration. Practical utility would come as a secondary benefit.

The new space agency, as imagined by the Purcell Committee, would face significant challenges in constructing and managing what promised to be the greatest technological undertaking yet. It would have to develop and build machines that could work in wholly alien environments without the benefit of a pilot on hand to continually monitor onboard systems, all the while solving the inevitable problems that would crop up. The agency would also have to serve some military needs without sacrificing the needs of the science community—while still reassuring the public its nation's defense, science, and technology were first in the world.

Even the military goals in space were facing similarly daunting challenges. Looking around at his colleagues, Killian saw that Sputnik had cast a spell over military officers, causing otherwise rational commanders to wax romantic about space as a realm for exploration and a battlefield in the next war. In both war and peace, staying on the cutting edge in this new realm was both paramount and alluring.

Though the recommendations from Washington called for restructuring the NACA, by the spring of 1958 it was still a bill waiting to become a law. In the meantime, the military branches pressed forward with their own spaceflight programs. For the air force, it was certain that if the new civilian agency did come to pass, it would rely on the service for just about everything, much like the NACA had. Taking the X-15 program as a model, the air force assumed the new agency would govern the space program and test the spacecraft but that the service would get the glory of flying the missions and completely lofty goals. The air force thus invited the NACA to undertake a joint study on spaceflight under the assumption that it would be an air force program, one that would be run by the Air Research and Development Command under General Bernhard Schriever. Extensive and ambitious, the air force had its sights firmly set on landing a man on the Moon as the final stage of an incremental program.

The first phase of the air force's Man in Space program was a simple, technical demonstration phase called Man in Space Soonest, or MISS, designed to take the initial steps in space and explore the human factors involved. The phase would begin with six robotic flights to test the hardware and flight systems, followed by six animal flights over the course of six months to test the life support system. Pending success on these missions, the first manned flight would launch as early as October 1960. These initial piloted missions would add reentry and recovery to the air force's knowledge base, the two key mission events that needed to be worked out before sending men on more complicated flights.

Because the goals of this phase were fairly basic, the spacecraft would be equivalently simple, a high-drag, zero-lift, blunt-nosed cylindrical vehicle eight feet in diameter with a flared bottom with an ablative heat shield on the Earth-facing surface. The pilot would lie on his back on a couch inside a pressurized cabin, though he would still wear a pressure suit for safety. The capsule's cabin would also house the main guidance and control systems, as well as the secondary power pack and a telemetry and voice communications system that would establish a link between the astronaut and ground crews. In the flared skirt of the spacecraft would be the reaction control jets, the retrorockets for reentry, and the recovery parachutes for a splashdown at sea.

The whole phase was designed to gain a better understanding of the human side of spaceflight. And because no one could be sure how a man would function in microgravity, these first flights would be almost entirely automated. It was possible that a pilot might be fine, or he might become disoriented and panic, rendering him a hazard to himself. But if the first flights were successful, and if the pilot was found capable of making decisions in space, the air force would give him increased control in subsequent flights, beginning with manual control over his attitude in space and retrofire burn.

The human factor goals of the MISS fed into the next phase, called Man in Space Sophisticated, or MISSOPH, a phase that was subdivided into three sections. Beginning in March of 1961, MISSOPH I would launch robotic and animal flights in larger spacecraft capable of staying aloft for up to two weeks, which was roughly the projected time for a round-trip mission to the Moon. The spacecraft for this phase would be more or less a larger version of the previous MISS version featuring a new airlock. Building on lessons learned in MISSOPH I, MISSOPH II would take advantage of the larger Super Titan Fluorine booster to put a much larger and more complex vehicle into a highly elliptical orbit reaching as high as forty thousand miles from the Earth. From this altitude, the spacecraft would reenter the atmosphere at a breakneck speed of thirty-five thousand feet per second or about 23,800 miles per hour, which is approximately the same speed as a return from the Moon. Having survived this harsh reentry, the MISSOPH II vehicle would become the prototype for a lunar return vehicle.

The third stage, MISSOPH III, would bring increased sophistication into the spacecraft. Unlike the earlier vehicles, this one would be specifically designed to give the pilot more control for precision landings. It wasn't a capsule; this vehicle would feature a flat triangular bottom reminiscent of a boost-glide vehicle. MISSOPH III would also debut the first spacesuit capable of supporting an astronaut leaving his spacecraft to work in the vacuum of space.

MISSOPH III would live beyond its specific phase, serving as an Earth-orbital vehicle as well as the circumlunar vehicle for lunar missions. But first, unmanned vehicles would scope out this distant world. Lunar Reconnaissance or LUREC was the air force's unmanned third phase intended to run simultaneously with the MISSOPH phase beginning April 1960. It was also subdivided into stages. LUREC I was devoted to figuring out the details of real-time tracking and communications with a spacecraft a quarter of a million miles away in the vicinity of the Moon. Once the tracking system was in place the LUREC II stage could launch. This stage was devoted to testing the guidance system that would support a flight to the Moon and gather data on the lunar environment in anticipation of a landing. Using an array of scientific instruments, these unmanned vehicles would measure the temperature, radioactivity, and atmospheric density around the Moon, sending back television images to help narrow down safe landing sites. Having gained a better understanding of the lunar environment, LUREC III would be the first to attempt a soft landing on the Moon's surface. Using rockets to slow its descent and telescoping legs to cushion the impact, this spacecraft would be the first to study the Moon up close, adding seismic and audio data from ground noises to our understanding of our natural satellite.

The robotic LUREC III flights would be one-way missions, but the manned spacecraft would follow in its wake. Manned Lunar Flight, or LUMAN, was the last phase of this air force program, and it would see a man landed on the Moon and returned safely to the Earth. The first stage, LUMAN I, called for animal flights around the Moon to verify the hardware, computer, and life support systems, a relatively simple mission expected to fly as early as May 1962. LUMAN II would see the same mission launched with men on board in place of animals. LUMAN III would see an unmanned spacecraft landed on the

Moon, while LUMAN IV would complete the goal of returning a spacecraft from the lunar surface, ideally early in 1963.

The manned landing would finally come with LUMAN V. Following the same profile as the two previous stages, an astronaut would pilot this spacecraft to a soft touchdown on the lunar surface. Once there, he would leave the spacecraft through the airlock and, thanks to his special pressure suit, leave the vehicle and explore the alien territory. With this mission, the program's main objective would be met. The subsequent LUMAN VI and LUMAN VII would see additional landed and orbital missions launched with increasingly sophisticated scientific instruments to better understand the Moon.

This air force vision for space exploration was certainly grand but not completely out of the realm of possibility. The Man in Space program was projected to cost $1.5 billion from the first unmanned missions through to the LUMAN missions slated to launch by 1965. But strict conditions would have to be met to keep the whole endeavor on schedule and under budget. The projected cost and time frame hinged on the program beginning on July 1, 1958, just months from the time this proposal reached air force headquarters. Success also demanded that all control of the program be given to the air force, which would in turn be free to consolidate whatever resources were needed and collaborate when beneficial.

An added incentive to pursuing this program, argued the air force, was the spinoff technologies. Improved reconnaissance, communications, and an early warning system for enemy attacks were three capabilities that promised to benefit the U.S. military, and they were all capabilities expected to come from developing the lunar landing program. The Man in Space program, the air force said, would not only restore America's national prestige, it would have an important psychological impact on the whole world. Landing a man on the Moon would without a doubt raise America back to a technologically dominant position in the eyes of the world.

However feasible, this lofty air force proposal was eventually scaled back to focus on the first stage, Man in Space Soonest. The first phase was one that could be done quickly and before taking on the challenge of sending a man to the Moon. But there were other similarly simplistic proposals floating around at the time to compete with MISS.

In the wake of the successful Manhigh II flight, the army's Wernher von Braun had contacted the air force's David Simons regarding the capsule. The German engineer saw it as the perfect vehicle for spaceflight, sophisticated enough to keep a man alive in orbit but also simple enough that it could be easily modified and launched on a rocket. Von Braun asked Simons whether he would be interested in developing a capsule along the line of Manhigh that could launch into space on a Redstone, turning the gondola and missile into a basic spaceflight system. Simons was on board; the two men saw eye to eye on the idea that capsules were the most effective means to get the first men into orbit. The fruits of this collaboration spawned a proposal von Braun aptly called Man Very High, and it was a program more or less the same as the air force's own MISS program. A simple manned spacecraft would launch into orbit on a Redstone, gathering data before considering grander goals like a flight to the Moon. The army, too, saw an American military presence on Earth's natural satellite as inevitable. This proposed joint project between the army and the air force promised fast results since the bulk of the technology already existed.

But interservice rivalry ultimately trumped any collaboration between von Braun and Simons; the air force and army couldn't decide who would get credit for the program if it worked. A resolute von Braun adapted his proposal to a purely army version called Project Adam as a reference to the biblical first man. For this program, the army would build its own capsule and launch it on one of von Braun's modified Redstones. General Medaris and the Army Ballistic Missile Association had begun a campaign to gain control over the nation's space program not long after successfully putting Explorer I into orbit.

Not to be left out, the navy was also eager to eke out its place in space. Von Braun had invited the navy to be a part of both Project Man Very High and Project Adam for the simple reason that the navy was best equipped to recover a spacecraft after it splashed down in the ocean. But the navy wanted to be more involved, and in April 1958 its Bureau of Aeronautics presented a manned satellite program to ARPA. Called MER for Manned Earth Reconnaissance, the proposal called for launching a manned spacecraft shaped like a cylinder with spherical ends. Once in orbit, the ends of the cylinder would expand laterally along two telescoping beams turning the pod into a delta-wing inflated glider with a rigid nose. The whole system would be pilot controlled from launch to gliding landing on the water, satisfying the requirements of an early spaceflight program.

These capsule-based military space proposals weren't the only ideas circulating that spring. The air force and NACA were still pursuing a boost-glide vehicle as the next step after the X-15. Based on Eugen Sänger's antipodal bomber system, the program was alternatively called BoMi for Bomber Missile, Brass Bell in its incarnation as a dedicated reconnaissance vehicle, and ROBO for Rocket Bomber. ROBO was the version ultimately investigated by industry partners Douglas, Convair, and North American Aviation. Still a futuristic technology, there was deemed no need to fast-track the program into development, and so the proposal called for an incremental development phase to bring the system to life. Called weapons system 464L, it was eventually nicknamed Dyna-Soar in reference to its dynamic soaring landing profile. It quickly became a case study of a developing program designed to create longevity behind America's first steps into space, along with a weapons system that would shape America's military arsenal, meeting military needs through to 1980.

The first phase spacecraft, Dyna-Soar I, would be the conceptual test article and the first vehicle that would go beyond the X-15 with a pilot on board. Air launched from a mother ship, this first iteration would reach altitudes above 250,000 feet and return the first hard data on hypersonic flight as well as aerodynamic, structural, heating, and human data during its short flights that would ultimately feed into the later versions. The next vehicle would be Dyna-Soar II, essentially the same version that had previously been called Brass Bell. This was to be a reconnaissance weapons system that would use a rocket engine to reach a peak altitude of just

170,000 feet but fly at speeds up to 12,300 miles per hour, gathering intelligence in the name of national security. The next iteration, Dyna-Soar III, was similar to the earlier ROBO concept, a hypersonic, global, strategic bombardment and reconnaissance weapon that would double as a manned multistage rocket-powered glider aircraft, capable of circumnavigating the globe.

The military space proposals were all pitched to the Department of Defense with the understanding that ARPA would be the governing body behind the nation's spaceflight program. But in the beginning of May, the military on the whole lost the fight for control in space when President Eisenhower's call for a civilian space agency was answered by Congress, who drew up the National Aeronautics and Space Act.

The National Aeronautics and Space (NAS) Act outlined exactly what this new civilian agency would do, largely according to Killian's stipulations. The act officially separated civilian and military sectors in space, solidifying policies in favor of the peaceful use of space. It also called for the absorption of the NACA as the core organization that would become the backbone of the new National Aeronautics and Space Administration, its expanse of laboratories and thousands of staff members included. There was some provision for NASA to absorb other pockets of space research in the country as well, including the Army Ballistic Missile Association, since their rockets would be integral to the future of launch vehicles. But even so, absorbing these military pockets, NASA, as Killian and Eisenhower had devised it, would remain a strictly civilian enterprise with very limited military participation. The president also recruited Senator Lyndon Johnson from Texas, a strong proponent of an American spaceflight program, to fight for the space agency's creation. It turned out to be a shrewd move on Eisenhower's part. Johnson created the Space Council that served to move the NAS Act through congressional challenges quickly, and just three months later, on July 29, the NAS Act was signed into law.

The next question was over who should lead the new NASA. Hugh Dryden was an obvious choice. He would bring nearly a decade of experience at the helm of the NACA, a career marked by striking foresight into aeronautical technologies, and a reputation of having a clear knowledge of the intricacies involved in running a series of national research laboratories to the new agency. But congressional leaders thought Dryden's style of directing the NACA quietly gave the impression that he was too laid-back, not an exuberant enough character to spearhead the nation's exciting move into space. Convinced that space would invigorate the nation, Congress wanted a leader as charismatic as the future was bright. The conversation led to another possible candidate: Thomas Keith Glennan was president of the Case Institute of Technology and a familiar face to many within the NACA.

With Eisenhower's approval, Killian set about to court Glennan into accepting the appointment, and though space was the new frontier, it wasn't an easy sell. The future of space and whether it would remain a viable industry after getting a man into orbit remained uncertain. Glennan had reservations and told Killian he needed to think over the offer. Days later, Glennan made his decision. He would accept the appointment as NASA's administrator, he told Killian, on the condition that Dryden serve as his deputy. Killian agreed, and Dryden accepted the appointment. Both Glennan and Dryden were formally appointed on August 8 and were sworn in eleven days later as the first two leaders of NASA.

With an administrator and his deputy in place, the NACA's transition to NASA began. The pace was rapid and the change not without its challenges. Turning a research community with a host of small programs into one multimillion-dollar organization running its own cutting-edge programs was a daunting task. NASA also prepared to absorb some of the military programs underway to place them firmly in the civilian space sphere. The air force and the Department of Defense space programs were poised to transfer to the new NASA, as were a handful of centers around the country, including the NACA Lewis, Langley, and Ames laboratories as well as the High Speed Flight Station. Talks began about moving von Braun and the Army Ballistic Missile Association, as well as

Caltech's Jet Propulsion Laboratory, under NASA's umbrella as well. Any center with any ties to space would fall under the civilian agency before long. It was in the nation's best interests for NASA to develop its own spaceflight program with the country's best minds and technologies readily on hand.

Toward the end of August, NACA employees around the country watched a video message from their leader. Dryden appeared on the screen, but instead of identifying himself as the NACA's director, he introduced himself to the eight thousand people under his employ as the deputy administrator of NASA. He then introduced Glennan, the man who would soon be their new leader.

We have come to a new day, Glennan began, before addressing the reality that the era of the NACA was fast coming to a close. Rather than feeling loss, he urged his future employees to take great pride in the coming change. Of all the agencies nationwide, he said, the NACA has been selected to have the honor of ushering in the new era of exploration on behalf of the country. It was the achievements of each and every NACA employee that gave this historic agency the reputation that left it as the emissary in space. "The NASA," he went on, would certainly be different than its predecessor owing to its different mission, but it would surely thrive under the same model of forward-thinking technical development that had served the NACA so well in its nearly half century of operation.

The NACA would never cease to exist, Glennan promised his staff. The transition was a positive one, a metamorphosis that would draw together the best the nation had to offer toward this exciting new goal. And that metamorphosis would be complete by the end of the workday on September 30.

Glennan was true to his word. On October 1, 1958, a Tuesday morning, thousands of NACA employees around the country went to work as they always did, only they weren't working strictly in aeronautics anymore. They were working for NASA, and the change in their agency's name was palpable. Working for the nation's space agency, the sky was no longer their limit.

Epilogue: America Finds Its Footing in Space

On October 7, 1958, with his agency one week old, NASA Administrator T. Keith Glennan considered proposals for a manned spaceflight program. One was for a ballistic capsule, the kind pioneered by Max Faget from the Langley Research Laboratory. It was shaped like a truncated cone with a rounded bottom over which was strapped a packet of retro-fire rockets designed to begin the capsule's fall from orbit. The upper portion of the capsule was an elongated neck, housing the parachute that would slow its final descent toward splashdown. In the name of catching up to the Russians in space, speed trumped sophistication; this was the fastest way to get a man in space first. Glennan gave the program his glib endorsement, saying simply, "let's get on with it." It became NASA's singular goal to get an American astronaut in space before the Soviet Union launched a cosmonaut.

The next day, the thirty-five-member Space Task Group was informally established at Langley Field in Virginia to bring the manned spaceflight program to fruition. In the weeks that followed, these men and women traveled around the country determining which existing programs would be absorbed by NASA to achieve the manned spaceflight goal. Some visited the air force's Wright Air Development Center in Ohio to learn about human factors in near space. Others studied different ablation materials for reentry heat shields alongside representatives from the Air Force Ballistic Missile Division. Engineers at Langley started working on the parachute that would slow this capsule during its final descent. Some Langley personnel met with Wernher von Braun's team at the Army Ballistic Missile Agency to discuss using the Redstone missile to launch suborbital missions. Others visited the Air Force Ballistic Missile Division to discuss using the Atlas missile to launch orbital flights.

The prospect of using spaceplanes to send pilots into orbit was all but forgotten when the doors of a large hangar opened on a sunny Wednesday morning on October 15 at North American Aviation's plant in Los Angeles. A small yellow tractor drove out from a hangar into the sunshine trailing a sleek black aircraft behind it flanked by two white-coated engineers. The first production X-15 made its way slowly through the parted crowd of gathered dignitaries and military personnel before coming to a stop in front of Vice President Richard Nixon who stood on a small stage set up in the shade. Nixon toured the sophisticated aircraft, designed to take men to the fringes of space that was now overshadowed by a simplistic capsule. By the end of October, specifications for NASA's ballistic spacecraft were firm enough for the agency to release a call for proposals. Industry partners were free to pitch their own versions of the vehicle, outlining the program and suggesting methods of analysis and construction.

NASA pressed on, as did its partner agencies. The air force's Atlas program reached a milestone distance flight of over six thousand miles, bringing the rocket one step closer to launching an astronaut. Test vehicles were developed for reentry assessments. Contractors submitted their proposals to the space agency, and small-scale rockets were developed with the intention of testing the spacecraft's abort system. During a December 17 policy speech, Glennan referred to the manned spaceflight effort for the first time by name, Project Mercury, solidifying the program in the minds of the American public.

In January 1959, NASA determined the physical characteristics of the men who would fly as part of the Mercury program. Hopeful astronauts had to be under forty years of age, shorter than five feet eleven inches, and weigh less than 180 pounds, physical constraints dictated by the size of the capsule and power of the Redstone and Atlas rockets. Candidates also had to be graduates of a test pilot school and have fifteen hundred hours' flying time in jet aircraft. Of the 508 men who met these basic requirements for NASA's astronaut program, 110 applied and began a rigorous screening process and series of extensive medical examinations. Meanwhile, further technical pieces of the puzzle fell into place. Abort systems and ablative materials were tested, instruments were designed and honed, and test trajectories were calculated. The U.S. Navy was the final military partner to join the program, signing on to support NASA in recovering the astronauts after splashdown.

Six months and nine days after the space agency formally opened its doors, NASA held a press conference at its new headquarters in Washington, D.C. Seated behind a long table between two American flags were seven men, all wearing dress shirts, dark jackets, and ties, two of which were bow ties. Behind them hung NASA's insignia, a blue circle representing a planet with the agency's acronym in white letters. Stars in the sphere's body were meant to signify space, a red chevron crossing the sphere denoting a wing as a callback to the new agency's ties to aeronautics, and a thin white ellipse showed an orbiting spacecraft. On the floor in front of the table were two models, one of an Atlas rocket with a Mercury capsule on top and the other a larger version of the spacecraft.

Glennan introduced the men at the table: Malcolm S. Carpenter, Leroy G. Cooper, John H. Glenn, Virgil I. Grissom, Walter M. Schirra, Alan B. Shepard, and Donald K. Slayton. "These, ladies and gentleman," he finished as the room broke into applause, "are the nation's Mercury astronauts." NASA and military officials spoke before the press was able to address the astronauts directly, asking after their family lives and selection process. Some of the pilots seemed somewhat uncomfortable in the spotlight, but soon won the room over.

A picture of the Mercury astronauts was featured the next day on the front page of the *New York Times*, and small articles with biographies of the men appeared in other newspapers and magazines throughout the country in the days that followed. The Mercury astronauts weren't headline news like Sputnik had been, but without having done anything except speak during a press conference they quickly became national heroes. The fear that had gripped the nation over uncertainty in space was replaced by optimism personified. The Soviet threat still loomed, but now with NASA, America had its answer.

Glossary of People

Neil Armstrong: August 5, 1930–August 25, 2012. Pilot engineer who joined the NACA in 1955 as a test pilot at the High Speed Flight Station. He flew some of the first missions with reaction controls mounted on an X-1B.

Magnus von Braun: May 10, 1919–June 21, 2003. Wernher von Braun's younger brother who found and surrendered the German rocket team to American soldiers at the end of the Second World War.

Wernher von Braun: March 23, 1912–June 16, 1977. One of the lead designers of the German A-4/V-2 program imported to the United States after the Second World War then moved to the U.S. Army to develop the Redstone family of rockets.

Albert Scott Crossfield: October 2, 1921–April 19, 2006. Pilot engineer who joined the NACA in 1950. He was the first man to reach Mach 2 and helped bring the X-15 rocket-powered aircraft to life at North American Aviation.

Walter Dornberger: September 6, 1895–June 27, 1980. Leader of the German Army's rocketry program that developed the A-4/V-2. After emigrating to the United States, Dornberger consulted for the U.S. Air Force before taking a job with Bell Aircraft where he pitched the concept of hypersonic ultra planes.

Hugh Dryden: July 2, 1898–December 2, 1965. Aerodynamicist with the National Bureau of Standards, member of the Army Air Force Science Advisory Group and the peacetime Science Advisory Board, and director of the National Advisory Committee for Aeronautics.

Dwight D. Eisenhower: October 14, 1890–March 28, 1969. American army general, president of the United States from 1953 to 1961. Eisenhower was instrumental in establishing NASA as a civilian agency rather than folding the space program within a military branch.

Theodore von Kármán: May 11, 1881–May 6, 1963. Hungarian mathematician, aerospace engineer and physicist, head of the Army Air Force Science Advisory Group and professor at the California Institute of Technology (Caltech) and founder of the Jet Propulsion Laboratory, which is now a NASA center.

Hermann Oberth: June 25, 1894–December 29, 1989. A father of rocketry whose *Die Rakete zu den Planetenräumen (The Rocket into Planetary Space)* inspired other pioneers in rocketry including von Braun and Valier.

David Simons: June 7, 1922–April 5, 2010. U.S. Air Force flight surgeon who launched early biological payloads on V-2 variants called Blossom rockets and piloted the Manhigh II balloon flight.

John Paul Stapp: July 11, 1910–November 13, 1999. U.S. Air Force flight surgeon known for his human deceleration tests, of which he was a frequent test subject, and pioneer investigating human tolerances in extreme environments.

Max Valier: February 9, 1895–May 17, 1930. Popularizer of rocketry in Germany known for attaching rockets to cars, sleds, and sailplanes. Founding member of the Verein für Raumschiffahrt (Society for Space Travel).

Glossary of Places and Organizations

Army Air Force: The flying branch of America's military service until 1947 when it was separated to become the standalone United States Air Force.

ARPA: The Advanced Research Projects Agency, created by Secretary of Defense Neil McElroy in 1958 to manage advanced programs such as spaceflight programs. In 1972, DARPA was established as a separate defense agency, then renamed ARPA in 1993, then changed back to DARPA in 1996.

High Speed Flight Station: The NACA's center at Edwards Air Force Base. The center was renamed the High Speed Flight Station in 1954, the Dryden Flight Research Center in 1976, and the Armstrong Flight Research Center in 2014.

Kummersdorf West: The German Army's rocket facility outside Berlin where Dornberger, von Braun, and their engineering colleagues developed the Aggregate series of rockets.

Mittelwerk: The underground factory in the Harz Mountains in Germany where A-4/V-2 rockets were built using concentration camp prisoner labor.

Muroc Air Force Base: Established in 1933 by Army Air Force Lieutenant Colonel Henry H. "Hap" Arnold as a bombing and gunnery range. In 1949 it was renamed Edwards Air Force Base in honor of Captain Glen Edwards.

NACA: The National Advisory Committee for Aeronautics. Pronunciations vary, but typically each letter is pronounced individually as N-A-C-A as opposed to "NASA," which is pronounced as a single word.

NACA Ames: The second NACA laboratory established in Sunnyvale, California, in 1939. It is named for physicist Joseph Sweetman Ames, one of the founding members of the NACA.

NACA Langley Memorial Laboratory: The first NACA laboratory established in 1917, named for American astronomer, physicist, and aviation pioneer Samuel Pierpont Langley. It is now a NASA center.

NACA Lewis Research Center: Established as an NACA center in 1942 as the Aircraft Engine Research Laboratory, it was renamed the Flight Propulsion Research Laboratory in 1947 then the

Lewis Flight Propulsion Laboratory in 1948 in honor of George W. Lewis, head of the NACA from 1919 to 1947. In 1999, the site was renamed the NASA John H. Glenn Research Center in honor of the first American in orbit.

Peenemünde: The German Army's rocket facility on the northern German island of Usedom where Dornberger, von Braun, and their rocket engineer colleagues built the A-4/V-2.

Raketenflugplatz: The Verein für Raumschiffahrt's rocket development site seventeen miles outside Berlin.

Redstone Arsenal: A World War II munitions production site closed after the war then reopened as the Ordnance Rocket Center in 1949 where it hosted the German scientists and their rocket research and development programs. The site was also home to the Army Ballistic Missile Agency from 1956 to 1960 when it was transferred to NASA's George C. Marshall Space Flight Center.

Verein für Raumschiffahrt: The VfR, or Society for Space Travel, founded by Max Valier. Notable members included Oberth and von Braun.

White Sands Proving Ground: A U.S. Army missile range in New Mexico where the first recovered V-2s were launched in America. In 1958, the site was renamed the White Sands Missile Range.

Glossary of Rockets

The Aggregate series: The German-built series of rockets that ultimately led to the V-2 offensive weapon.

A-1: The first in the series featured a rear-mounted engine. It had a one-foot diameter and was 4.6 feet long.

A-2: The second A-rocket featured a gyroscope in the center of the rocket's body, a change the team hoped would solve the instability problems of the A-1 with the same dimensions. The rockets Max and Moritz were A-2s.

A-3: The third A-rocket stood 21.3 feet tall and measured 2.3 feet around at its widest point and featured small stabilizing fins.

A-4: Also known as the V-2 (Vergeltungswaffe Zwei), this 47-foot tall rocket that measured 5.5 feet in diameter was the German Army's first operational missile.

A-5: Though numerically after the A-4, this rocket actually preceded the first operational rocket. It was an intermediary stage to rectify problems with the A-3.

A-6: Sixth in the Aggregate series, this rocket was a concept study designed to test different propellants.

A-7: This Aggregate rocket was a concept study featuring small wings. Unlike its predecessors, it was designed to launch from underneath an aircraft on high arcing trajectories rather than launch from an upright position on the ground.

A-8: This eighth Aggregate rocket was designed with a longer fuselage than any predecessors.

A-9: The ninth in the Aggregate series, this was a concept for an A-4 with wide, swept-back wings running from its nose to midsection turning the rocket into a glider that could coast through the atmosphere rather than arc over it. One conceptual variant included a pressurized cockpit for a pilot on board. One A-9 was launched in 1945 under the name A-4b.

A-10: The last in the Aggregate series, this was another concept and the first for a multistage weapons system. It consisted of an A-9 stacked on top of an 85-ton booster. The two-stage configuration could have covered the distance between a western European launch site and a major city on America's East Coast.

Atlas: Originally designated MX-1593, this U.S. Air Force intercontinental ballistic missile first flew in 1957.

Hermes: Hermes was the American V-2 program designed to understand the German missile and eventually create an American offshoot.

Hermes A1: This was planned as an antiaircraft missile.

Hermes A2: A surface-to-surface missile.

Hermes A3: This was designed to deliver a 1,000-pound warhead over a distance of 150 miles with an error of just 200 feet.

Hermes II: This was a Hermes variant designed to use a ramjet engine, a type of engine that literally rams air into the combustion chamber without any moving parts.

Hermes C1: A three-stage missile that used clusters of solid fuel rockets to generate enough power to deliver larger payloads to distant targets.

Jupiter: A longer-range offshoot of the Redstone rocket.

Jupiter A: The first generation Jupiter missile, Jupiter A was designed to gather test data of the guidance system as well as evolved separation procedures for multistage missiles.

Jupiter C: Three modified Redstones were designated Jupiter C for composite reentry vehicles that would test ablative materials on scale nose cones.

Kegeldüse: A basic engine designed by Hermann Oberth featuring a hollow steel cone as a combustion chamber.

Mirak: Short for *minimumrakete* meaning "simple rocket," the Miraks looked like firecrackers: a simple copper rocket engine similar to the Kegeldüse engine in a cylindrical fuselage sitting behind a bullet-shaped cover. A long aluminum tube served as a guiding stick.

Redstone: The Hermes C1 was renamed for the Redstone Arsenal on April 8, 1952.

Repulsor: Designed in early 1931, this rocket had a liquid oxygen and gasoline-fueled engine encased in water, and featured rear fins for stability and a long support stick for guidance.

Thor: An air force intermediate-range ballistic missile based on components of the Atlas missile.

Titan: The United States's first real multistage intercontinental ballistic missile that started flying at the end of the 1950s.

Selected Notes

Chapter 1: Hobby Rocketeers

11   at the end of a warm, clear Saturday: Record of Climatological Observations, National Climatic Data Center.

11   he pushed for one final test: Essers, *Max Valier: A Pioneer of Space Travel*, 210.

15   Unwilling to stand by and be ridiculed: Essers, *Max Valier: A Pioneer of Space Travel*, 147.

17   Stamer managed one of the finest landings of his career: Stamer's recollection on the flight appears in Essers, *Max Valier: A Pioneer of Space Travel*, 158.

18   "Help to create the spaceship!": Essers, *Max Valier: A Pioneer of Space Travel*, 170

18   most felt Valier's showmanship denigrated what they were trying to do: Essers, *Max Valier: A Pioneer of Space Travel*, 184.

23   just ten German marks: Piszkiewicz, *The Nazi Rocketeers: Dreams of Space and Crimes of War*, 12.

23   The Raketenflugplatz slowly took shape: Piszkiewicz, *The Nazi Rocketeers: Dreams of Space and Crimes of War*, 14.

25   happy resolution for the VfR, and an entertaining newsreel: Piszkiewicz, *The Nazi Rocketeers: Dreams of Space and Crimes of War*, 16.

Chapter 2: The Rocket Loophole

29   The army representatives weren't impressed with the VfR: Ordway and Sharpe, *The Rocket Team*, 18–19.

30   The sophistication of Kummersdorf West awed the VfR pioneers: Piszkiewicz, *The Nazi Rocketeers: Dreams of Space and Crimes of War*, 18.

30   Wernher von Braun did leave a mark on Dornberger: Stuhlinger and Ordway, *Wernher von Braun: Crusader for Space*, 215.

30   Both men knew a partnership would be beneficial: Piszkiewicz, *The Nazi Rocketeers: Dreams of Space and Crimes of War*, 22–23.

30   not everyone at the Raketenflugplatz shared his enthusiasm: Neufeld, *Von Braun: Dreamer of Space, Engineer of War*, 54–55.

30   they wanted to build exploratory rockets, not missiles: Piszkiewicz, *The Nazi Rocketeers: Dreams of Space and Crimes of War*, 22.

35   the solution came through the von Brauns: Neufeld, *Von Braun: Dreamer of Space, Engineer of War*, 80.

36   Sänger imagined a future: Myhra, *Sänger: Germany's Orbital Rocket Bomber in WWII*, 49.

38   The Austrian army thought it unlikely: Myhra, *Sänger: Germany's Orbital Rocket Bomber in WWII*, 55.

39   The Luftwaffe had no problem with Sänger's Austrian background: Myhra, *Sänger: Germany's Orbital Rocket Bomber in WWII*, 61.

40   came with the caveat that he join the Nazi Party: Neufeld, *Von Braun: Dreamer of Space, Engineer of War*, 90.

41   parachute was the most immediate cause: Neufeld, *Von Braun: Dreamer of Space, Engineer of War*, 104.

43   His eyes, thought Dornberger, seemed unfocused: Piszkiewicz, *The Nazi Rocketeers: Dreams of Space and Crimes of War*, 53.

46   he worried that his American forces wouldn't be ready: Smith, *Eisenhower in War and Peace*, 167.

49   Dornberger could feel the tension in the air rise: Dornberger, *V-2*, 16.

49   the spaceship, Dornberger knew, had been born with that launch: Dornberger, *V-2*, 25.

49   Himmler was there to learn as much as he could about the weapon: Piszkiewicz, *The Nazi Rocketeers: Dreams of Space and Crimes of War*, 75–76.

50   the group retired to the Hearth Room: Neufeld, *Von Braun: Dreamer of Space, Engineer of War*, 146.

51 The Führer was impressed, far more than he had been after his 1939 visit: Dornberger, *V-2*, 96–97.

Chapter 3: The Turning Tide of War

53 It was a familiar sound, as were the sounds of planes overhead: Neufeld, *Von Braun: Dreamer of Space, Engineer of War*, 153–55.

53 slid on his bedroom slippers: Dornberger, *V-2*, 141.

56 Quality control remained an issue as more A-4s came out of Mittelwerk: Neufeld, *The Rocket and the Reich: Peenemünde and the Coming of the Ballistic Missile Era*, 225.

57 persistent knocking in the early hours of the morning: Neufeld, *Von Braun: Dreamer of Space, Engineer of War*, 169.

58 Dornberger knew all four men were indispensable to the A-4 effort: Dornberger, *V-2*, 179.

58 strong-arm control of the rocket program: Piszkiewicz, *The Nazi Rocketeers: Dreams of Space and Crimes of War*, 111.

60 Roosevelt's indecision lasted the full five days: Smith, *Eisenhower in War and Peace*, 316.

62 they also assumed it was a feint: Smith, *Eisenhower in War and Peace*, 362.

65 expert opinions on how these technologies might benefit the future: "Memorandum for Dr. von Karman" in von Kármán, "Towards New Horizons," iii.

68 turned to their leader for guidance: Piszkiewicz, *The Nazi Rocketeers: Dreams of Space and Crimes of War*, 179.

69 the V-2 scientists and sites being high on everyone's list: Bob Ward, *Dr. Space: The Life of Wernher von Braun*, 55–6.

69 the rocket team could hear artillery fire: Neufeld, *Von Braun: Dreamer of Space, Engineer of War*, 190.

70 Conflicting orders gave him some freedom to pick: Neufeld, *Von Braun: Dreamer of Space, Engineer of War*, 192.

Chapter 4: Escape and Surrender

72   It was one of the few times, if not the only time, he exploited this title: Neufeld, *Von Braun: Dreamer of Space, Engineer of War*, 193.

73   by raising his left arm above his head: Neufeld, *Von Braun: Dreamer of Space, Engineer of War*, 195.

74   It was one of Eisenhower's happy moments in the war: Smith, *Eisenhower in War and Peace*, 420.

76   watched the soldier's face as he pictured the scenario: Piszkiewicz, *The Nazi Rocketeers: Dreams of Space and Crimes of War*, 199.

79   and was the most expendable: Correspondence between Magnus von Braun and Francis French, September 8, 1995.

80   recognized the bounty that had fallen into their hands and became quite friendly: Correspondence between Magnus von Braun and Francis French, September 8, 1995.

82   Celebrity, it seemed, suited von Braun: Neufeld, *Von Braun: Dreamer of Space, Engineer of War*, 201.

Chapter 5: Nazi Rockets in New Mexico

89   Once home to hunters and agricultural villages: Eidenbach, "Cultural History of the Tularosa Basin," http://www.nps.gov/whsa/historyculture/cultural-history-of-the-tularosa-basin.htm.

92   The group recognized that nations on both sides had begun: Von Kármán, "Where We Stand: A Report of the AAF Scientific Advisory Board," iv.

92   von Kármán similarly considered them the most capable missile research group: Von Kármán, "Where We Stand: A Report of the AAF Scientific Advisory Board," 13.

94   if Hitler had given the Peenemünde team priority status and more support: Von Kármán, "Where We Stand: A Report of the AAF Scientific Advisory Board," 13.

94   Dryden offered his opinion that the next major conflict:
       Dryden, et al., *Guided Missiles and Pilotless Aircraft: A
       Report of the AAF Scientific Advisory Board*, 1.

94   if this technological trend continued, the next major
       conflict: Von Kármán, "Towards New Horizons," xi.

95   his reception was far from warm: Neufeld, *Von Braun:
       Dreamer of Space, Engineer of War*, 215.

102   developing this technology into a viable missile now,
       he believed: Eisenhower, *Waging Peace*, 207.

103   subtext was that they might not be needed: Brzezinski,
       Red Moon Rising, 88.

105   he went to the American consulate: Neufeld, *Von
       Braun: Dreamer of Space, Engineer of War*, 245.

Chapter 6: Rockets Meet Airplanes

108   Through a string of profanities and abuse: Arnold,
       Global Mission, 136–37.

109   the variables including the heating properties of air and
       density: Mack ed., *From Engineering Science to Big Science*,
       62.

112   Bob Woods stopped by Ezra Kotcher's office: Mack ed.,
       From Engineering Science to Big Science, 86.

113   Kotcher was sure the only way to break the sound barrier:
       Mack ed., *From Engineering Science to Big Science*, 87.

116   with the gentle precision of a surgeon: *Time*, "Army &
       Navy: What Comes Naturally." Monday, Dec. 23, 1946.

117   various flight test engineers refer to the X-1 as a death
       trap: Yeager and Janos, *Yeager*, 119.

118   Boyd considered Yeager to be the most naturally
       instinctive pilot: Yeager and Janos, *Yeager*, 126.

118   Colonel Boyd told Yeager it was his: Yeager and Janos,
       Yeager, 123.

119   likely landed in a wastepaper basket: Crossfield and
       Blair Jr., *Always Another Dawn: The Story of a Rocket Test
       Pilot*, 23.

119   Yeager's wife, who had not been happy: Yeager and
       Janos, *Yeager*, 163.

120   he privately promised, he would push the airplane through the sound barrier: Yeager and Janos, *Yeager*, 164.

120   without his ears or anything else falling off: Yeager and Janos, *Yeager*, 165.

Chapter 7: A New War, a New Missile, and a New Leader

124   Crossfield was disappointed: Crossfield and Blair Jr., *Always Another Dawn*, 28–29.

124   a pioneering spirit among the small group of men: Crossfield and Blair Jr., *Always Another Dawn*, 30.

125   Truman had asked Army General Dwight D. Eisenhower to command: Smith, *Eisenhower in War and Peace*, 493.

126   was willing to seek another term in office: Smith, *Eisenhower in War and Peace*, 504.

127   If he was offered the Republican nomination for the presidency: Smith, *Eisenhower in War and Peace*, 510.

127   Eisenhower realized he hadn't been so emotional in years: Smith, *Eisenhower in War and Peace*, 511–12.

Chapter 8: Higher and Faster

135   Dornberger he would be tried in his stead for the crime of launching rockets: Ordway III and Sharpe, *The Rocket Team*, 303.

136   Bell, Dornberger alluded to his former colleague … von Braun spent sleepless nights: Neufeld, *Von Braun: Dreamer of Space, Engineer of War*, 300.

136   Dornberger found a sympathetic and willing collaborator in Robert Woods: Jenkins and Landis, *Hypersonics: The Story of the North American X-15*, 11.

137   Dornberger reasoned that rocket propulsion would follow a similar path: Godwin, *Dyna-Soar: Hypersonic Strategic Weapon System*, 24.

140 By making the world more accessible, he anticipated, more people: Godwin, *Dyna-Soar: Hypersonic Strategic Weapon System*, 250.

141 Scott Crossfield knew he could reach Mach 2 in the Douglas D-558-II Skyrocket: Crossfield and Blair Jr., *Always Another Dawn*, 161.

142 a navy pilot, Dryden dictated, who would be the one to push the airplane: Crossfield and Blair Jr., *Always Another Dawn*, 168.

142 Crossfield offered to make an attempt at reaching Mach 2: Crossfield and Blair Jr., *Always Another Dawn*, 168.

142 far more concerned with the Skyrocket's health than his own: Crossfield and Blair Jr., *Always Another Dawn*, 171.

144 he at least wanted to break Crossfield's record: Yeager and Janos, *Yeager*, 250.

145 Yeager realized he was flying too fast at too high an altitude: Yeager and Janos, *Yeager*, 253.

146 Crossfield was, Dryden countered, needed at Edwards: Crossfield and Blair Jr., *Always Another Dawn*, 152.

148 the B-36, was too unknown to the team at the High Speed Flight Station: Crossfield and Blair Jr., *Always Another Dawn*, 264.

151 To Crossfield, North American was the default winner: Crossfield and Blair Jr., *Always Another Dawn*, 205.

152 North American turned out not to be too keen on the idea: Jenkins and Landis, *Hypersonics: The Story of the North American X-15*, 100–01.

152 One man, however, was extremely anxious to secure the X-15 contract: Jenkins and Landis, *Hypersonics: The Story of the North American X-15*, 101.

152 in the midst of the X-15 contract discussions, Rice called Storms: Crossfield and Blair Jr., *Always Another Dawn*, 205.

153 Scott Crossfield had some lingering concerns: Crossfield and Blair Jr., *Always Another Dawn*, 206.

153   He told Atwood he was a man who could bring a valuable: Crossfield and Blair Jr., *Always Another Dawn*, 210.

154   To Neil Armstrong, also a young pilot engineer: Hansen, *First Man: The Life of Neil A. Armstrong*, 120.

Chapter 9: Edging into Hypersonics

155   Scott Crossfield pulled his car into a parking lot: Crossfield and Blair Jr., *Always Another Dawn*, 217.

156   Feltz had never heard of the hypersonic research plane until: Crossfield and Blair Jr., *Always Another Dawn*, 220.

157   gave himself an unofficial role as "the X-15's chief son-of-a-bitch": Crossfield and Blair Jr., *Always Another Dawn*, 224–25.

158   Feltz came up with an elegant solution to increase the aircraft's lift: Crossfield and Blair Jr., *Always Another Dawn*, 229.

159   Storms descended on the team in a manner befitting his last name: Crossfield and Blair Jr., *Always Another Dawn*, 248–49.

160   Feltz agreed with Crossfield. In an emergency: Crossfield and Blair Jr., *Always Another Dawn*, 231.

162   Crossfield strongly preferred the Navy's Clark-designed: Crossfield and Blair Jr., *Always Another Dawn*, 240.

164   Walt Williams was eager to take over the aircraft: Crossfield and Blair Jr., *Always Another Dawn*, 213.

165   He knew from studying previous flights that the X-2: Merlin, "Starbursters: 55 Years Ago Capt. Mel Apt Conquered Mach 3, Lost Life on Fated Flight."

167   But it was exactly this type of flying that appealed to Armstrong: Hansen, *First Man: The Life of Neil A. Armstrong*, 122.

Chapter 10: The Floating Astronaut

171   Stapp quickly gained a reputation for operating: Ryan, *The Pre-Astronauts: Manned Ballooning on the Threshold of Space*, 16.

174   Kittinger watched as the sled flew across the desert: Kittinger and Ryan, *Come Up and Get Me*, 44.

174   Immobilized by his restrains, Stapp felt unbearable pain: Ryan, *The Pre-Astronauts: Manned Ballooning on the Threshold of Space*, 29.

175   he was already planning to add more rockets to Sonic Wind No. 1: Kittinger and Ryan, *Come Up and Get Me*, 45.

175   became a research niche for Stapp: Ryan, *The Pre-Astronauts: Manned Ballooning on the Threshold of Space*, 15.

176   Aircraft, Stapp knew, wouldn't be a suitable means: *The Pre-Astronauts: Manned Ballooning on the Threshold of Space*, 19.

177   Simons also saw balloons as the best option: Kennedy, *Touching Space*, 45.

178   Stapp walked into Simons's office wondering: Ryan, *The Pre-Astronauts*, 23.

179   Stapp asked if Simons would be willing to make the flight: Ryan, *The Pre-Astronauts*, 24.

183   He wanted this first flight to go to a test pilot: Ryan, *The Pre-Astronauts*, 65.

183   For Kittinger, the very precise flight profile: Kittinger and Ryan, *Come Up and Get Me*, 41.

184   light refused to illuminate did Simons start to become concerned: Ryan, *The Pre-Astronauts*, 31.

185   Kittinger asked thousands of questions: Kittinger and Ryan, *Come Up and Get Me*, 48.

186   Simons finally confronted Stapp: Ryan, *The Pre-Astronauts*, 78.

186   Kittinger, for his part, approached his Manhigh flight: Kittinger and Ryan, *Come Up and Get Me*, 48.

186   Kittinger took matters into his own hands: Kittinger and Ryan, *Come Up and Get Me*, 61.

188   Not only did he want to complete the test: Kittinger and Ryan, *Come Up and Get Me*, 63.

189   looking out the portholes he saw the balloon: Kittinger and Ryan, *Come Up and Get Me*, 63.

189   Simons became increasingly anxious as the mission wore on: Kittinger and Ryan, *Come Up and Get Me*, 65.

190   he toyed with Simons just a little: Kittinger and Ryan, *Come Up and Get Me*, 66.

Chapter 11: Space Becomes an Option

195   he called the IGY a unique and striking example: Eisenhower, letter to Chester I. Barnard, June 24, 1954.

197   was clear to von Braun that multistage rockets: Neufeld, *Von Braun: Dreamer of Space, Engineer of War*, 279.

198   interest in the talks of spaceflight outweighed the skepticism: Davey, "San Antonio & the Genesis of the Collier's Series, 'Man Will Conquer Space Soon!' 54.

198   Ryan saw von Braun again as he was leaving another session: Davey, "San Antonio & the Genesis of the Collier's Series, 'Man Will Conquer Space Soon!' 55.

208   Vanguard alone promoted the idealistic notion that science: Killian, *Sputnik, Scientists, and Eisenhower: A Memoir of the First Special Assistant to the President for Science and Technology*, 119.

209   When von Braun heard the decision: Neufeld, *Von Braun: Dreamer of Space, Engineer of War*, 295.

209   scrambling to revise the Orbiter proposal: Neufeld, *Von Braun: Dreamer of Space, Engineer of War*, 298.

Chapter 12: The First Satellite Race

211   The visit was an unscheduled stop: Neufeld, *Von Braun: Dreamer of Space, Engineer of War*, 298.

213   von Braun felt slightly vindicated: Neufeld, *Von Braun: Dreamer of Space, Engineer of War*, 299.

215   Pentagon sent an official to Cape Canaveral: Neufeld, *Von Braun: Dreamer of Space, Engineer of War*, 304.

215   He and his team were forbidden from discussing: Neufeld, *Von Braun: Dreamer of Space, Engineer of War*, 304.

217 Schriever supported satellite development on the condition: USAF, "Proposal for Man-in-Space (1957–1958)," 30.

223 The most important result of the International Geophysical Year: Hagerty, "Remarks by the President in Connection with the Opening of the International Geophysical Year," June 30, 1957.

Chapter 13: One Little Ball's Big Impact

225 Wernher von Braun ducked briefly back to his office: Neufeld, *Von Braun: Dreamer of Space, Engineer of War*, 311.

225 voice on the other end ask what he thought about it: Neufeld, *Von Braun: Dreamer of Space, Engineer of War*, 311–12.

225 The news didn't entirely shock von Braun: Killian, *Sputnik, Scientists, and Eisenhower*, 2.

226 Medaris watched as the frustration of Project Orbiter: Medaris and Gordon, *Countdown for Decision*, 155.

226 his meal marked by a constant stream of arguments: Medaris and Gordon, *Countdown for Decision*, 155.

226 Medaris told von Braun to take Redstone RS-29: Neufeld, *Von Braun: Dreamer of Space, Engineer of War*, 312.

227–8 Detractors seized the event as proof of Eisenhower's failure: Smith, *Eisenhower in War and Peace*, 732.

229 Yes, the army could have launched a satellite: Killian, *Sputnik, Scientists, and Eisenhower*, 3.

229 He worried more than anything that Sputnik: Killian, *Sputnik, Scientists, and Eisenhower*, 7.

229 The president also ordered outgoing Secretary Wilson: Eisenhower, *Waging Peace*, 211–12.

230 Yes, he admitted, the United States could have orbited: "Official White House Transcript of President Eisenhower's Press and Radio Conference #123." October 9, 1957.

232 The consensus was that American science had not: Killian, *Sputnik, Scientists, and Eisenhower*, 15.

232 offered the president of the Polaroid company, Dr. E. H. Land: Killian, *Sputnik, Scientists, and Eisenhower*, 16.

232 Dr. Rabi had a suggestion: Killian, *Sputnik, Scientists, and Eisenhower*, 16.

232 asked whether Killian would be willing to travel: Killian, *Sputnik, Scientists, and Eisenhower*, 20.

233 And the president's willingness to take council: Killian, *Sputnik, Scientists, and Eisenhower*, 29.

235 The same basic principle could be applied to a return: Hartman, *Adventures in Research*, 263.

236 Lighting a cigarette before getting into his car: Gray, *Angle of Attack*, 41.

236 he took the proposal to Washington: Jenkins, *X-15: Extending the Frontiers of Flight*, 330.

238 My subject tonight is science and national security: Eisenhower, "Radio and Television Address to the American People on Science in National Security," November 7, 1957.

239 Foremost, he said, was to bring together: Eisenhower, "Radio and Television Address to the American People on Science in National Security," November 7, 1957.

240 On the whole, Killian considered this group: Killian, *Sputnik, Scientists, and Eisenhower*, 109.

240 The committee's conclusions were synthesized: USAF, "Proposal for Man-in-Space (1957–1958)," 72.

244 remained unwavering in his conviction that America: Kistiakowsky, "Memorandum for Dr. Killian," December 19, 1957.

249 Von Braun had to wait for tracking stations to pick up: Neufeld, *Von Braun: Dreamer of Space, Engineer of War*, 323.

Chapter 14: The Fight to Control Space

251 President Dwight Eisenhower was privately relieved: Eisenhower, *Waging Peace*, 255–57.

252 Vera Winzen volunteered to climb a ladder: Ryan, *The Pre-Astronauts: Manned Ballooning on the Threshold of Space*, 100.

257   Nothing included in the statement was science fiction: Eisenhower, "President's Science Advisory Committee, Introduction to Outer Space," March 26, 1958.

257   Killian saw that Sputnik had cast a spell over military: Killian, *Sputnik, Scientists, and Eisenhower*, 128.

262   the army's Wernher von Braun had contacted the air force's David Simons: Ryan, *The Pre-Astronauts: Manned Ballooning on the Threshold of Space*, 149.

265   Killian set about to court Glennan: Killian, *Sputnik, Scientists, and Eisenhower*, 139.

266   We have come to a new day, Glennan began: "Creation of NASA: Message to Employees of NACA from T. Keith Glennan 1958 NASA." YouTube via Jeff Quitney.

Epilogue: America Finds Its Footing in Space

267   Glennan gave the program his glib endorsement: Swenson Jr. et al., *This New Ocean*, 109.

268   Nixon toured the sophisticated aircraft: "1958.10.15 X-15 ROLLOUT AND POEM TEST PILOT" by user "Jim Davis."

269   "These, ladies and gentleman," he finished: "Press Conference Introducing the Mercury Astronauts," NASA video.

Bibliography

"Army & Navy: What Comes Naturally." *Time*. Monday, December 23, 1946.

Arnold, H. H. 1949. *Global Mission*. Harper & Brothers, New York.

Berger, Carl. 1966. "The Air Force in Space Fiscal Year 1962." USAF Historical Division Liaison Office.

"Big Maneuvers Test U.S. Army" *Life Magazine* vol. 11 no. 14. October 6, 1941.

von Braun, Magnus, letter to Francis French, September 8, 1995.

von Braun, Magnus, letter to Francis French, March 8, 1996.

von Braun, Magnus, letter to Francis French, May 29, 1996.

von Braun, Magnus, letter to Francis French, August 5, 1996.

Brzezinski, Matthew. 2007. *Red Moon Rising*. Times Books, New York.

Bullard, John W. 1965. *History of the Redstone Missile System*. Army Missile Command, Washington, D.C.

"Chronology of Early Air Force Man-in-Space Activity 1955–1960." Historical Division, Office of Information Space Systems Division, Air Force Systems Command, United States Air Force.

Cleary, Mark C. *Army Ballistic Missile Programs at Cape Canaveral 1953–1988*. 45th Space Wing History Office.

"Creation of NASA: Message to Employees of NACA from T. Keith Glennan 1958 NASA," posted by user "Jeff Quitney," https://www.youtube.com/watch?v=ic4G-8qX_bk (May 17, 2013).

Crossfield, A. Scott, and Clay Blair Jr., 1960. *Always Another Dawn: The Story of a Rocket Test Pilot*. The World Publishing Company, Cleveland.

D'Antonia, Michael. 2007. *A Dog, a Ball, and a Monkey*. Simon and Schuster, New York.

Davey, Colin. "San Antonio & the Genesis of the Collier's Series, 'Man Will Conquer Space Soon!' " *Horizons*, March/April 2013, 54–55.

"Development of the Corporal: The Embryo of the Army Missile Program." 1961. Army Missile Command, Huntsville.

Dickson, Paul. 2001. *Sputnik: The Shock of the Century*. Berkeley Publishing Group, New York.

Dobbs, Michael. 2012. *Six Months in 1945*. Knopf, New York.

Dryden, Hugh L. et al. 1946. *Guided Missiles and Pilotless Aircraft: A Report Prepared for the AAF Scientific Advisory Group.* Headquarters Air Materiel Command, Dayton.

Dryden, Hugh L. et al. 1946.*Guidance and Homing of Missiles and Pilotless Aircraft: A Report Prepared for the AAF Scientific Advisory Group.* Headquarters Air Materiel Command, Dayton.

Dryden, Hugh L. 1965. *Theodore von Kármán.* National Academy of Sciences, Washington.

Dykman, J. T. *Fifty Years Ago: Winter of Discontent, Winter 1951–52.* The Eisenhower Institute, Gettysburg.

Eidenbach, Peter L. "Cultural History of the Tularosa Basin." National Parks Service. http://www.nps.gov/whsa/learn/historyculture/cultural-history-of-the-tularosa-basin.htm (July 9, 2015).

Eisenhower, Dwight D. 1948. *Crusade in Europe.* Johns Hopkins University Press, Baltimore.

Eisenhower, Dwight D., letter to Chester I. Barnard, June 24, 1954.

Eisenhower, Dwight D. "Radio and Television Address to the American People on Science in National Security," transcript, the American Presidency Project. November 7, 1957.

Eisenhower, Dwight D. 1965. *Waging Peace: The White House Years 1956–1961.* Doubleday & Company, Garden City.

Essers, I. 1976. *Max Valier: A Pioneer of Space Travel.* National Aeronautics and Space Administration, Washington, D.C.

Evans, Michelle. 2013. *The X-15 Rocket Plane: Flying the First Wings into Space.* University of Nebraska Press, Lincoln.

"Explorer-I and Jupiter-C." Data Sheet, Department of Astronautics, National Air and Space Museum, Smithsonian Institution, Washington, D.C.

Ezell, Edward Clinton, and Linda Neuman Ezell. 1978. *The Partnership: A History of the Apollo-Soyuz Test Project.* National Aeronautics and Space Administration, Washington, D.C.

Foerstner, Abigail. 2007. *James Van Allen: The First Eight Billion Miles.* University of Iowa Press, Iowa City.

Gabel, Christopher R. 1991. *The US Army GHQ Maneuvers of 1941.* United States Army, Washington, D.C.

Gainor, Chris. 2008. *To a Distant Day: The Rocket Pioneer.* University of Nebraska Press, Lincoln.

Godwin, Robert, ed. 2003. *Dyna-Soar: Hypersonic Strategic Weapon System.* Apogee, Burlington.

Godwin, Robert, ed. 2000. *X-15: The NASA Mission Reports.* Apogee, Burlington.

Gorn, Michael H. 1992. *The Universal Man: Theodore von Kármán's Life in Aeronautics*. Smithsonian, Washington, D.C.

Gray, Mike. 1992. *Angle of Attack: Harrison Storms and the Race to the Moon*. W. W. Norton & Company, New York.

Grimwood, James M., and Frances Strowd. 1962. *History of the Jupiter Missile System*. U.S. Army Ordnance Missile Command.

Hagerty, John C. White House Press Statement. July 29, 1957.

Hagerty, John C. "Remarks by the President in Connection with the Opening of the International Geophysical Year." White House Press Statement. June 30, 1957.

Hallion, Richard P. 1984. *On the Frontier: Flight Research at Dryden, 1946–1981*. National Aeronautics and Space Administration, Washington, D.C.

Hansen, James R. 1987. *Engineer in Charge: A History of the Langley Aeronautical Laboratory, 1917–1958*. National Aeronautics and Space Administration.

Hansen, James. 2005. *First Man: The Life of Neil A. Armstrong*. Simon & Schuster, New York.

Hartman, Edwin P. *Adventures in Research: A History of Ames Research Center 1940–1965*. National Aeronautics and Space Administration, Washington, D.C.

"History of Research in Space Biology and Biodynamics at the U.S. Air Force Missile Development Center, Holloman Air Force Base, New Mexico, 1946–1958." Historical Division. Office of Information Services, Air Force Missile Development Center, Air Research and Development Command, Holloman Air Force Base, New Mexico.

"History of Research in Space Biology and Biodynamics." 1958. Holloman Air Force Base, New Mexico.

"History of Strategic Air and Ballistic Missile Defense, Volume II, 1956–1972." http://www.history.army.mil/html/books/bmd/BMDV2.pdf (June 23, 2015).

Jenkins, Dennis R. 2000. *Hypersonics Before the Shuttle*. National Aeronautics and Space Administration, Washington, D.C.

Jenkins, Dennis R., and Tony R Landis. 2003. *Hypersonic: The Story of the North American X-15*. Specialty Press, North Branch.

Jenkins, Dennis R. 2007. *X-15: Extending the Frontiers of Flight*. National Aeronautics and Space Administration, Washington, D.C.

Kennedy, Gregory. 2007. *Touching Space: The Story of Project Manhigh*. Schiffer Military History, Atglen.

Kennedy, Gregory P. 2009. *The Rockets and Missiles of White Sands Proving Ground 1945–1948.* Schiffer Military History, Atglen.

Killian Jr., James R. 1977. *Sputnik, Scientists, and Eisenhower: A Memoir of the First Special Assistant to the President for Science and Technology.* MIT Press, Cambridge.

King, Benjamin, and Timothy Kutta. 1998. *Impact: The History of Germany's V-Weapons in World War II.* Sharpedon, Rockville Center.

Kistiakozsky, Piore, and York. "Memorandum for Dr. Killian." December 17, 1957.

Kittinger, Joe, and Craig Ryan. 2010. *Come Up and Get Me.* University of New Mexico Press, Albuquerque.

Laursen, V. "The Second International Polar Year (1932/33)." World Meteorological Organization. https://www.wmo.int/pages/mediacentre/documents/SecondInt.PolarYear.pdf (April 5, 2015).

Lee, Ernest "Tex." 1985. *Papers.* Dwight D. Eisenhower Library Online. Abilene.

Ley, Willy. "For Your Information," *Galaxy Magazine,* October 1955. https://archive.org/stream/galaxymagazine-1955-10/Galaxy_1955_10#page/n61/mode/2up (April 5, 2015).

Lundgren, William R. 1955. *Across the High Frontier: The Story of a Test Pilot—Major Charles E. Yeager, USAF.* Bantam Books, Toronto.

Mack, Pamela E., ed. 1998. *From Engineering Science to Big Science: The NACA and NASA Collier Trophy Research Project Winners.* National Aeronautics and Space Administration, Washington, D.C.

Manucy, Albert C. 1949. *Artillery Through the Ages.* National Parks Service Interpretive Theory, Washington, D.C.

McLaughlin Green, Constance, and Milton Lomask. 1970. *Vanguard: A History.* National Aeronautics and Space Administration, Washington, D.C.

Medaris, General J. B., and Arthur Gordon. 1960. *Countdown for Decision.* G. P. Putnam's Sons, New York.

Merlin, Peter W., "Starbuster: 55 Years Ago Capt. Mel Apt Conquered Mach 3, Lost Life on Fated Flight," Edwards Air Force Base, October 5, 2011, http://www.edwards.af.mil/news/story.asp?id=123274801.

Middlebrook, Martin. 1982. *The Peenemünde Raid.* Pen & Sword Books, Great Britain.

Mieczkowski, Yanek. 2013. *Eisenhower's Sputnik Moment: The Race for Space and World Prestige.* Cornell University Press, Ithaca.

Myhra, David. 2002. *Sänger: Germany's Orbital Rocket Bomber in World War II*. Schiffer Military History, Atglen.

Neufeld, Jacob. 1990. "The Development of Ballistic Missiles in the United States Air Force 1945–1960." United States Air Force, Washington, D.C.

Neufeld, Jacob. 2005. *Bernard A. Schriever: Challenging the Unknown*. Office of Air Force History, Washington, D.C.

Neufeld, Michael J. 1995. *The Rocket and the Reich: Peenemünde and the Coming of the Ballistic Missile Era*. The Free Press, New York.

Neufeld, Michael J. 2007. *Von Braun: Dreamer of Space, Engineer of War*. Alfred A. Knopf, New York.

Nicolet, M. "The International Geophysical Year 1957/58." World Meteorological Organization. https://www.wmo.int/pages/media center/documents/Int.GeophysicalYear.pdf (April 5, 2015).

"Official White House Transcript of President Eisenhower's Press and Radio Conference #123." October 9, 1957.

"Oral History: Maxime A. Faget." 1997. Interviewed by Jim Slade. Johnson Space Center, Houston.

Ordway III, Frederick I., and Mitchell R Sharpe. 1982. *The Rocket Team: From the V-2 to the Saturn Moon Rocket*. MIT Press, Cambridge.

Overy, R. J. 2004. *The Dictators: Hitler's Germany and Stalin's Russia*. W. W. Norton, New York.

Piszkiewicz, Dennis. 1995. *The Nazi Rocketeers: Dreams of Space and Crimes of War*. Stackpole Books, Mechanicsburg.

Powell-Willhite, Irene, ed. 1971. *The Voice of Dr. Wernher von Braun: An Anthology*. Apogee Books, Burlington.

"President's Science Advisory Committee, Introduction to Outer Space," March 26, 1958, pp. 1-2, 6, 13-15. NASA Historical Reference Colleciton, NASA History Division, NASA Headquarters, Washington, D.C.

"Proceedings of the X-15 First Flight 30th Anniversary Celebration." 1989. NASA Ames Research Center, Dryden Flight Research Facility, Edwards.

"Project Horizon: Volume I Summary and Supporting Considerations." 1959. United States Army, Washington, D.C.

"Proposal for Man-in-Space (1957–1958)." Air Research and Development Command, United States Air Force.

"Proposed United States Program for the International Geophysical Year 1957–1958." 1956. National Academy of Sciences National Research Council, Washington, D.C.

Reiffel, L. 1959. "A Study of Lunar Research Flights, Vol 1." Air Force Special Weapons Center, Kirkland Air Force Base, New Mexico.

Rickman, Gregg, ed. 2004. *The Science Fiction Film Reader Paperback.* Limelight Editions, New York.

Ryan, Craig. 1995. *The Pre-Astronauts: Manned Ballooning on the Threshold of Space.* Naval Institute Press, Annapolis.

Sänger, Eugen, and Irene Bredt. M. Hamermesh, trans. 1944. "A Rocket Drive for Long Range Bombers." Technical Information Branch Buaer Navy Department.

Satterfield, Paul H., and David S. Akens. 1958. "Historical Monograph: Army Ordnance Satellite Program." Army Ordnance Corps, Huntsville.

Sheppard, F. H. W. 1973. "The Crown Estate in Kensington Palace Gardens: Historical Development." *Survey of London.* Volume 37, Northern Kensington, ed. pp. 151–62.

Shirer, William L. 1961, 2011. *The Rise and Fall of the Third Reich.* Rosetta Books, New York.

Siddiqi, Asif A. "Korolev, Sputnik, and the International Geophysical Year." NASA. http://history.nasa.gov/sputnik/siddiqi.html (April 5, 2015).

Singer, S. F. 1954. "Studies of a Minimum Orbital Unmanned Satellite of the Earth. (Mouse)." American Rocket Society, New York.

Smith, Jean Edward. 2012. *Eisenhower in War and Peace.* Random House, New York.

"Soviet Fires New Satellite, Carrying Dog; Half-Ton Sphere Is Reported 900 Miles Up." *New York Times.* November 3, 1957. pp. 1.

"Soviet Fires Earth Satellite into Space: It Is Circling the Globe at 18,000 M.P.H.; Sphere Tracked in 4 Crossings Over U.S." *New York Times.* October 5, 1957. pp. 1.

Stewart, H. J. et al. 1945–1946. "Early Military Characteristics Expected of the Drawing-Board Corporal and the Estimated Performance of the ORDCIT Corporal Series of Guided Missiles." Jet Propulsion Laboratory, Pasadena.

Stuhlinger, Ernst, and Frederick I Ordway III. 1996. *Wernher von Braun: Crusader for Space.* Krieger Publishing Company, Malabar.

Sturm, Thomas A. 1967. *The USAF Scientific Advisory Board: Its First Twenty Years 1944–1964.* Office of Air Force History, Washington, D.C.

Swenson Jr., Loyd S., James M. Grimwood, and Charles C. Alexander. *This New Ocean: A History of Project Mercury.* National Aeronautics and Space Administration, Washington, D.C.

Thompson, Milton O. 1992. *At the Edge of Space.* Smithsonian, Washington, D.C.

Tregaskis, Richard. 1961. *X-15 Diary: The Story of America's First Space Ship.* Bison Books, University of Nebraska Press, Lincoln.

Tsien, H. S. et al. 1946. "Technical Intelligence Supplement: A Report of the AAF Scientific Advisory Board." Headquarters Air Materiel Command, Dayton.

Van Pelt, Michel. 2012. *Rocketing into the Future: The History and Technology of Rocket Planes.* Springer Praxis, Chichester.

Von Kármán, Theodore. 1946. "Where We Stand: A Report of the AAF Scientific Advisory Board." Headquarters Air Materiel Command, Dayton.

Von Kármán, Theodore. 1946. "Towards New Horizons: A Report to General of the Army H. H. Arnold Submitted on Behalf of the A.A.F. Scientific Advisory Group." Headquarters Air Materiel Command, Dayton.

Von Kármán, Theodore, and Lee Edson. 1967. *The Wind and Beyond: Theodore von Kármán, Pioneer in Aviation and Pathfinder in Space.* Little, Brown and Company, Boston.

"Walt Disney Treasures Tomorrowland: Disney in Space and Beyond," Walt Disney Studios Home Entertainment. DVD release May 18, 2004.

Ward, Bob. 2005. *Dr. Space: The Life of Wernher von Braun.* Naval Institute Press, Annapolis.

Wolfe, Allen E., and William J. Truscott. 1960. "Juno I: Re-Entry Test Vehicles and Explorer Satellites." Jet Propulsion Laboratory, Pasadena.

White, L. D. 1952. "Final Report: Project Hermes V-2 Missile Program." General Electric, Schenectady.

Winzen Staff. 1959. "Manhigh I." Air Force Missile Development Center, Alamogordo.

Yeager, Chuck, and Leo Janos. 1985. *Yeager.* Bantam Books, Toronto.

"1958.10.15 X-15 ROLLOUT AND POEM TEST PILOT" posted by user "Jim Davis," https://www.youtube.com/watch?v=4RHs9WYnnFM (April 2, 2008).

"1958 NASA/USAF Space Probes (Able-1) Final Report." Headquarters ARDC, Space Technology Laboratory, Los Angeles.

"7 Named as Pilots for Space Flights Scheduled in 1961." *New York Times.* April 10, 1959.

Acknowledgments

There are so many people without whom this book could not have happened. First and foremost, Jim Martin, whose e-mail I opened on my phone at the bank one day. You assured me yours was a real publishing offer and stood behind me every step of the way. Thank you, along with Jackie Johnson, Laura Phillips, and Bloomsbury, for reaching out to and guiding an unknown author.

I wasn't completely new to the science writing world when I started *Breaking the Chains of Gravity*. A number of people have helped me along in my career, getting me to the point where I was ready to take on a book. Alex Pasternack at *Motherboard*, thank you for giving me my first job as a writer. Fraser Cain at *Universe Today*, thank you for bringing me into the core of online space nerds, and for letting me come back from time to time even though I've been too busy for regular Hangouts. Thank you Ian O'Neill at *Discovery News* for hiring me and always having my back.

Two people reached out to help me enter the world of publishing, giving me advice I didn't even know I needed. Phil Marino at W. W. Norton Liveright, thank you for taking me to lunch more than two years ago in New York and giving me a publisher's opinion on an early proposal that ultimately turned into this book. Alex Jacobs from Cheney Literary, thank you for letting me ask you all kinds of questions about signing a book contract.

Francis French, your ongoing support, not to mention the opportunity to write a section of an Apollo astronaut's memoir, has meant the world as I continue to establish my career. Matt Wood, you've been a wonderful friend and a sympathetic ear through the final stages of this endeavor, and you're a great sounding board. Robert Reeves, you've been a source of strength on rougher days. Thank you, all three of you, for giving your time and thoughts on my early drafts.

A handful of people might not realize that they've been pretty important to me. Phil Plait and Geoff Notkin, you both offered me a lot of support years ago that helped give me confidence to keep going in this weird world of science communicating. Jamey Wetmore at ASU, thank you for helping me with my master's thesis all those summers ago; I'm still using lessons learned from you in my big writing projects. Nathan Kitada at YouTube, thank you for seeing something worthwhile in my space history videos; you've made me feel like I really can branch beyond the niche of already dedicated space history fans. Alan Stern, thank you for bringing me onto the New Horizons team, an incredible opportunity and an even more meaningful vote of confidence.

Mark Ulett, thank you for cheering me on when I first took on this project, for letting me bounce ideas off you for years as I honed this story, and for reading every early draft of my sample chapter and proposal.

To my parents, thank you for the gift of my education, your love and support, and the freedom to devote myself to this book. And Dad, for learning more than you ever wanted to know about spaceflight reading drafts along the way. I love you both.

And finally, to every Twitter follower, Facebook fan, YouTube subscriber, and blog reader. You are all the reason I'm able to do what I love for a living. Thank you for allowing me to explore my nerdy space passion and share it with you. I'm so glad you guys love this stuff as much as I do!

Amy Shira Teitel, 2015

Index

A-1 (Aggregate–1) and A–2
  (Max and Moritz) 32–34, 35
A-3 34, 35, 39–42, 43, 44
A-4 42, 43, 44, 48–51, 62–64, 93
  A-4b 67–68, 93
  concentration camp
    labour 55–56, 63, 85
  V–2 attacks on London 64, 68
A-5 42, 43, 44, 93
A-6 93
A-7 93
A-8 93
A-9 93
A-10 93
Adams, Sherman 232–33
Aerobee sounding rockets 193, 207
Aggregate rockets 32, 40, 92–93
Air Research and Development
    Command (ARDC) 244–46
  Man in Space program 258–62
aircraft testing 108–109
Alamogordo Army Air
    Field 89–90, 100
Allen, Harvey 234–35
America see USA
Ames Research Laboratory,
    California 123, 233, 234
Anglo-American Combined
    Intelligence Objectives
    Subcommittee 83
antipodal bombers 38–39, 64, 86,
    136–37, 141, 218, 234, 263
Apollo 8
Apt, Mel 164–66, 170
Armstrong, Neil 8, 154, 166–70
Army Air Force Scientific Advisory
    Group 65–67, 69, 74, 78,
    91–92, 95
Army Ballistic Missile Agency
    (ABMA) 212, 215–16, 225–26,
    262, 264–65, 267
Arnold, Henry H. "Hap" 65–66,
    92, 94–95, 107–108
astronauts 268–70
Atlas missiles 131–32, 212, 222,
    252, 254, 267–69

Atomic Energy Commission
    132, 256
Atwood, Lee 153
Austria 38–39, 42
AVCO 222, 245–46
Axster, Hebert 80

ballistic missiles 93
balloon flights 176–81, 184–90,
    193–94, 252–54
Bat guided missile 66
Becker, Carl 27–31, 36
Bell Aircraft Corporation 108,
    112–19, 136, 145–46, 150–51,
    164, 245
Berlin Institute of Technology 21
Berlin-Charlottenburg Technische
    Universität 27
Blossom 100–101, 177–78
Boyd, Albert 117–18
Bradley, Omar 74
Braun, Magnus Freiherr von 22
Braun, Magnus von 58, 77,
    79–81
Braun, Sigismund von 59
Braun, Wernher von 21–22,
    30–32, 35–37, 39–44, 265, 267
  A-4 launches 48–51
  escape from Nazi
    Germany 68–73, 75–77
  Huntsville, Alabama 104–106,
    129, 196, 198
  Man Very High 262
  Mars 102, 200–203
  Project Orbiter 197–98, 209–10,
    211–12
  RAF raid on
    Peenemünde 53–55
  Redstone missiles 136, 196,
    213–16
  satellite program 196–203,
    225–26, 239, 243–44, 248–49
  SS 44–45, 57–59, 85
  transfer to US 79, 81–83, 85–87
  V–2 66
  White Sands 95–96

Bredt, Irene 64, 86, 136
Britain 64, 68, 83, 90–91
Bromley, William 84
Bumper rockets 196–97

California Institute of
    Technology 65
Caltech 146, 152, 266
Cape Canaveral, Florida 214–15,
    229, 237, 243, 248
Carpenter, Malcolm S. 269
Chrysler 129–30, 196
Churchill, Winston 60, 69
Clark, David 162
Collier's 198–99, 202
Combined Services Detailed
    Interrogation Center 134–35
compressibility 109–10
Convair 96–97, 102, 207, 222,
    245, 263
Cooper, Leroy G. 269
Copernicus, Nicolaus 191
Crossfield, Scott 118–19, 123–24,
    141–44, 146, 150–51, 153–54,
    155–58, 159, 162–63, 169–70

D-558-II Skyrocket 141–44
Disney, Walt 199–200
    "Man in Space" 200, 205
dogs 237–38
Dornberger, Walter 27–28,
    30–36, 39–45, 62–64, 71, 86,
    93, 157, 234
    A-4 launches 48–51, 62–64
    emigration to USA 135–38
    escape from Nazi
        Germany 75–77
    interrogation by the
        British 134–35
    RAF raid on
        Peenemünde 53–55
    surrender to US 79–81, 83
    ultra planes 140–41, 158, 218
    von Braun's arrest 57–59
Douglas Aircraft 141, 150–52,
    160, 263
Dryden, Hugh 65–66, 69, 79, 82,
    94–95, 109, 119, 123, 142,
    146–47, 150, 152, 240, 247, 256,
    264–66
Dyna-Soar 263–64

Edwards Air Force Base 123–24,
    132, 141–42, 144–46, 148–49,
    153–55, 159, 163–66, 168,
    172–73, 176, 181, 189, 233,
    237, 256
Eggers, Alfred 234–35
Eisenhower, Dwight 47–48,
    60–61, 67, 69, 74, 209, 223,
    264–65
    International Geophysical Year
        (IGY) 195–96
    missile program 102–103
    Open Skies Treaty 204–205, 229
    presidential election 125–30
    space policy 254–57
    Sputnik 227–33, 238–40,
        244, 251
    Suez Crisis 216–17
Expendable Earth Orbiter 219–20
Expendable Lunar Vehicle,
    Pass-By 221
Explorer 1 244, 249, 251–52,
    255, 262

Faget, Max 235, 267
Feltz, Charles 153, 156–58, 160,
    162, 236
Fokker 27–28
Ford Instrument Company 130
Frau im Mond 19–21, 22, 24

g-forces 139, 160–61, 171–72,
    165–66, 171–72
Galileo Galilei 191
Garmisch-Partenkirchen 82–83,
    85, 134–35
Gee-Whizz 172–73
General Electric 85, 90, 98, 102,
    245, 251
Germany 8
    First World War 27–28
    Nazis take power 33–34
    Second World War 44–46, 67,
        73–74
Gestapo 49, 57, 59, 72, 78
Gilruth, Robert R. 114
Glenn, John H. 269
Glennan, Thomas Keith 265–69
Goddard, Robert 9
Goebbels, Joseph 64
Goodlin, Chalmers "Slick" 116–19

Göring Institute   39
Göring, Hermann   34, 39, 46
Grissom, Virgil I.   269
Gröttrup, Helmut   58
Grunow, Heinrich   31

Hagen, John P.   229, 231
Hamill, James   84, 95
Harbou, Thea von   19
Henry, James P.   100
Hermes A1, A2, A3   98
Hermes C1   98, 129
Hermes II   98–99
High Speed Flight Station   166–69
Hilton, W. F.   111
Himmler, Heinrich   45, 49–50, 55,
    57–58, 62–63, 71
Hitler, Adolf   33–34, 42–44, 46,
    48–51, 55–57, 62–63, 74, 79,
    94, 103
Holaday, William M.   229, 231
Holloman Air Force Base   100,
    173, 176, 178, 180, 183–84, 194
Horstig, Ritter von   27–28
Huntsville, Alabama   104–106, 129,
    196, 198, 211–12
Huzel, Dieter   75–76, 80
hydrogen bomb   130
hypersonic flight   45, 123, 136,
    140–41, 146–53
    X-15   155–63, 233–34

International Geophysical Year
    (IGY)   191–94, 195–96, 198,
    203–206, 208–209, 211, 223,
    229–30, 239, 241, 248
International Harvester
    Company   161

J. F. Eisfeld Powder and
    Pyrotechnical Works   16
Jet Propulsion Laboratory   146,
    197, 206, 211, 266
jet-powered aircraft   109, 151–52
Johnson, Lyndon   255, 264
Johnson, Roy W.   251–52
Jupiter missiles   130–31, 212–16,
    239, 244, 248, 251

Kammler, Hans   55–57, 62–63, 70,
    72–73, 75–76, 78, 135

Kaplan, Joseph   206
Keech, Ray   15–16
Kegeldüse   22, 29
Kennedy, John, F.   8
Kensington Palace Gardens,
    London   133–34
Kesselring, Albert   36
Killian, James   130–31, 229,
    232–33, 239–40, 244, 255–57,
    264–65
Kittinger, Joe   174, 183–90, 252
Korean War   104, 106, 124–25,
    128–29, 150–51, 167–68,
    204
Kotcher, Ezra   112–13
Krueger, Walter   47
Kummer, Major   76–77
Kummersdorf West   29–35, 39–41,
    43, 50–51, 92

Land, E. H.   232
Lang, Fritz   19–20, 21, 24, 56
Langley Memorial Laboratory   109,
    112, 115, 123–24, 149–51, 154,
    234–37
Lear, Ben   47
Lee, Ernest R. "Tex"   47
Ley, Willy   20, 199–200
Lindenberg, Hans   80
Lippisch, Alexander   17
Loki rockets   197, 206, 208, 211
Luftwaffe   34–36, 39

Mach, Ernst   110
Man in Space program   258–62
    LUMAN   260–61
    LUREC   260
    MISS   258–59
    MISSOPH   259–60
Manhigh   180–90, 252–54, 262
Manned Earth
    Reconnaissance   263
Manned Recoverable Earth
    Orbiter   220
Mars   102, 200–203
Marshall, George C.   46–48, 60
Martin Company   102, 206, 208,
    210, 217, 222, 243, 245
Massachusetts Institute of
    Technology   100, 130,
    229, 232

McElroy, Neil   225–27, 229,
    251, 255
McMath, Francis Charles   208–209
Me-163   113
Me-262   109
Me163B Comet   67
Medaris, John Bruce   212–15,
    225–27, 248, 262
Messerschmitt   67, 109, 113
Metropolis   19, 56
mice   178
Mirak 1 and 2
Mittelwerk   56–57, 63, 70, 84, 90
monkeys   100–101, 176–77
Moon   8, 83, 162, 199–201, 218,
    221–22, 257, 262
Muroc, California   107–108,
    116–17, 119, 121, 123–24, 142,
    154, 171
MX-774 1593   131
MX-774   96–97, 102, 114, 131

NACA   66, 109, 112–16, 119,
    123–24, 136, 141–42, 147–54,
    158, 164, 169, 172–73, 218, 234,
    247, 263
    civilian agency   255–58
    NASA   264–66
NASA   8–9, 264–70
National Advisory Committee for
    Aeronautics see NACA
National Aeronautics and Space
    (NAS) Act (USA)   264
National Aeronautics and Space
    Administration see NASA
National Security Act 1947
    (USA)   100
NATO (North Atlantic Treaty
    Organization)   104, 125–27, 216
Navajo missile   151
Naval Technical Mission   82
Nazi Party   33–34, 40
    Kristallnacht   42
Nebel, Rudolf   20–24, 28, 30
New York Times   227, 237, 269
Newton's third law   13
Nimwegen, Erich   71–72
Nixon, Richard   128, 268
North American Aviation   129–30,
    150–52, 218, 263
    X-15   155–56, 170

Northrop Corporation   113, 123,
    171, 173

Oberth, Hermann   12–14, 18–22,
    29, 36
    Die Rakete zu den
        Planetenräumen   13–14, 18,
        21–22
    Ways to Spaceflight   19
Opel, Fritz von   14–15
Open Skies Treaty   204–205, 229
Operation LUSTY   78–79,
    82–83
Operation Overcast   86–87, 103,
    105, 134
Operation Overlord   60–62

Pearl Harbor   47, 152, 227
Peenemünde, Germany
    35–36, 39–42, 44–45,
    48–51, 53–58
    RAF raid   53–55
Perkins, Oliver "Perk"   142
Polar Year   192
Polaris missiles   215
pressure suits   161–62
Project 1115 132, 205, 208
Project Adam   262
Project Feedback   132, 207–208,
    217
Project Hermes   69, 74, 84–85,
    90–91, 96–98, 102–103,
    105–106, 206
Project HYWARDS   234, 245
Project Mercury   268–70
Project Orbiter   197–98, 203,
    206–14, 217, 225–26, 229
Project Paperclip   103, 105, 134
Project Vanguard   206–11, 214,
    217, 225–26, 228–31, 238–39,
    242–45, 248
Purcell, Edward   257

Quarles, Don   131, 205–206, 208,
    229

Rabi, Isidor   232
RAF   53–55
Raketenflugplatz Berlin   23–25,
    27–30, 32–33, 37
Reaction Motors   96, 113–14

Redstone Arsenal, Alabama   105,
    129–31, 136, 196–97, 200,
    208–209, 211–15, 225–26,
    244, 248, 262, 267–68
Reich Institute for Chemistry and
    Technology   21–22
Republic Aviation   151
Repulsor   24, 29–30
Reynolds Metals   130
Ridley, Jack   120
Riedel, Klaus   21–23, 30, 40, 58
Riedel, Walter   11–12, 31–32, 40
ROBO   263
rocket cars
    Eisfeld-Valier Rak. 1   16
    Opel-Sander Rak. 1   14–15
    Opel-Sander Rak. 2   15
    Rak Bob   16
rocket planes
    Lippisch Ente   17–18
rockets, early   90–91
Roosevelt, Franklin D.   46,
    60–61, 69
Round Three Conference
    233–34, 236
RS-27   211–13, 215–16, 229
RS-29   211–12, 214, 226, 248
Rudolph, Arthur   11, 12, 31, 40
Ryan, Cornelius   198–99

SAG   65–66, 69, 92–95
Sander, Friedrich Wilhelm   14, 17
Sänger, Eugen   36–39, 45, 93
    antipodal bomber   38–39, 64, 86,
    136–37, 263
satellites   131–32, 136, 194–99,
    202–207, 209–15, 217, 219,
    221–23
    Explorer 1   244, 249, 251–52,
    255, 262
    Sputnik   225–34, 236–38,
    240–41, 248, 251, 257, 269
Schirra, Walter M.   269
Schriever, Bernhard   217, 258
Second World War   8
Sergeant rockets   146, 206, 211, 244
Shephard, Alan B.   269
Siemens Company   21
Simons, David   99–101, 176–90,
    194, 252–54, 262
Slayton, Donald K.   269

Smith, Walter Bedell
Sonic Wind No. 1   173–75
Soviet Union   103–104, 130,
    196, 203–205, 207, 209–10,
    213, 216
Spaatz, Carl A. 95
space program   218–21
    Air Research and Development
    Command (ARDC)
    244–46
spaceflight   83, 131, 146, 247–48
Speer, Albert   48, 59
Sputnik   225–34, 236–38, 240–41,
    248, 251, 257, 269
SS (Schutzstaffel)   38, 44–45, 50, 57,
    70, 71–72, 76, 85
SS-6 Sapwood missile   216
Stack, John   109, 112–13
Stalin, Joseph   60, 68–69, 86, 104
Stamer, Friedrich   17–18
Stapp, John Paul   171–76, 178–81,
    183–87, 190, 194, 253
Staver, Robert   84
Stevenson, Adlai   128, 217
Stewart, Charles   80–81
Stewart, Homer Joe   206–209,
    211–12, 217, 229
Storms, Harrison   152–53, 156, 159.
    236–37
Stratolab   180, 197
Suez Crisis   216–17
supersonic flight   40, 48, 64,
    92–94, 102
    manned aircraft   110–14,
    117–18, 120–21, 123, 136,
    140, 146, 148, 151, 156,
    233–34, 245
    ultra planes   137–41

Taifun missiles   197
Technical University, Vienna   36
Teller, Edward   240–41
Tessmann, Bernhard   75–76, 80
Thor missiles   131, 212–13
Titan missiles   131, 210
Tofty, Holger   84, 86, 95
Treaty of Versailles   28
Truman, Harry S.   125–26
Tsiolkovsky, Konstantin   9
TV-2   229–30
TV-3   230–31, 242

ultra planes   137–41, 158, 218
United Kingdom *see* Britain
Universum Film-Aktien
    Gesellschaft   20–21, 24
Unmanned Recoverable Earth
    Orbiter   219
USA   8–9, 64–65
    Army Air Force Scientific
        Advisory Group   65–66
    preparing for war   46–48

V-2   64–65, 66, 69–70, 75, 82–85,
    151, 200
    Bumper rockets   196–97
    US V-2 program   74, 83–85,
        90–106, 208
    V-2 attacks on London   64, 68
    V-2 Upper Atmospheric Research
        Panel   97
Valier, Max   11–19, 22–24, 36, 200
Verein für Raumschiffahrt
    (VfR)   18–25, 28–33, 40, 199
Verne, Jules *De la Terre à la
    Lune*   12
Viking   102, 206–208
Volkhart, Kurt C.   14
von Hindenburg, Paul   33
von Kármán, Theodore   65, 74, 79,
    82, 92, 94–95, 109, 152
von Richthofen, Manfred   34
von Richthofen, Wolfram   34–36

WAC Corporal   197
Weyprecht, Carl   191–92

White Sands, New Mexico   89–91,
    95–99, 102, 104, 106
Williams, Walter C.   115–16,
    123–24, 142, 153, 164, 169–70
Wilson, Charles   131, 212, 216,
    225, 229
Winzen, Otto   177–81, 185–89,
    252–53
Winzen, Vera   178, 186–87,
    189, 252
Woods, Bob   112–13, 136, 141,
    146
Woolams, P. V. "Jack"   115
Wright Air Development
    Center   219, 221, 267
Wright Field, Ohio   100, 108, 112,
    115, 117–18, 172, 175

X-1   113–21, 124, 141, 151
X-15   149–53, 155–63, 169–70,
    176, 218, 233–34, 236–37, 245,
    247, 255, 258, 263, 268
X-15B   236–37, 247
X-1A   141–42, 144–45
X-1B   169
X-2   146, 150–51, 153, 163–66,
    170
XLR-99   156–57, 158
XP-59A Airacomet   108
XP-79
YB-49 Flying Wing   123
Yeager, Chuck   117–21, 141–42,
    144–45, 148–49, 160, 166